Advances in Intelligent and Soft Computing

94

Editor-in-Chief: J. Kacprzyk

T0137746

Advances in Intelligent and Soft Computing

Editor-in-Chief

Prof. Janusz Kacprzyk
Systems Research Institute
Polish Academy of Sciences
ul. Newelska 6
01-447 Warsaw
Poland
E-mail: kacprzyk@ibspan.waw.pl

Further volumes of this series can be found on our homepage: springer.com

José M. Molina,
José Ramón Casar Corredera,
Manuel Felipe Cátedra Pérez,
Javier Ortega-García, and
Ana M. Bernardos Barbolla (Eds.)

User-Centric Technologies and Applications

Proceedings of the CONTEXTS 2011
Workshop

 Springer

Editors

Prof. José M. Molina
Universidad Carlos III de Madrid
Departamento de Informática
Escuela Politécnica Superior
(campus de Colmenarejo)
Avda. Universidad Carlos III, 22
28270 Colmenarejo (MADRID)
Spain

Prof. José Ramón Casar Corredera
Universidad Politécnica de Madrid
Departamento de Señales, Sistemas
y Radiocomunicaciones
ETSI Telecomunicación
Av. Complutense 30
28040 Madrid
Spain

Prof. Manuel Felipe Cátedra Pérez
Universidad de Alcalá
Departamento de Ciencias
de la Computación
Escuela Politécnica
28806 Alcalá de Henares, Madrid
Spain

Prof. Javier Ortega-García
Universidad Autónoma de Madrid
Departamento de Tecnología
Electrónica y de las Comunicaciones
Escuela Politécnica Superior
c/ Francisco Tomás y Valiente, 11
28049 Madrid
Spain

Prof. Ana M. Bernardos Barbolla
Universidad Politécnica de Madrid
Departamento de Señales, Sistemas
y Radiocomunicaciones
ETSI Telecomunicación
Av. Complutense 30
28040 Madrid
Spain

ISBN 978-3-642-19907-3   e-ISBN 978-3-642-19908-0

DOI 10.1007/978-3-642-19908-0

Advances in Intelligent and Soft Computing   ISSN 1867-5662

Library of Congress Control Number: 2011923542

Typeset & Cover Design: Scientific Publishing Services Pvt. Ltd., Chennai, India.

Printed on acid-free paper
5 4 3 2 1 0
springer.com

Preface

The idea of living, working or relaxing in fully sensitive and connected environments led at the end of the eighties to the exploration of technologies to design ubiquitous 'user-centric' applications. Later, Schilit and Theimer in 1994 coined the term 'context-awareness' (sensibility or adaptation to the context) to describe how ubiquitous applications could react and adapt themselves to the changes of the environment and the user's situation. Context was then defined as 'any information that can be used to characterize the situation of an entity' (A. Dey, 1999).

Nowadays, the increasing availability of sensor networks, the miniaturization and the progressive reduction of sensors' price and the penetration of advanced mobile personal devices make possible to collect and process data from multiple sources and combine them in order to provide the applications with a sufficiently good Quality of Information about the user's context.

Thus, the CONTEXTS Workshop aims at gathering advances on key elements that enable the design and deployment of personalized and ubiquitous context-aware services and uses natural non-intrusive methods for interaction. In this Volume, the reader will find contributions on the following topics:

1. Tools to facilitated the design, deployment and operation of the context measurement elements and networks.
2. Sensor fusion methods to intelligently process context data: techniques for positioning with emerging technologies, cooperative management of the user-mobile-environment interaction and distributed context inference.
3. Context representation and management: data models to represent context information, methods to reason with uncertain information and techniques to manage the quality of service of context-aware systems.
4. Techniques oriented to identification and personalization, including biometry and other techniques with no-cooperative sensors like cameras or RF measurement devices.

The Workshop is organized and funded by the CONTEXTS Programme, one of the largest cooperative research initiatives in Madrid, with the participation of institutions from all over Spain and with the support of international experts. This programme focuses on advancing the key elements in frontier communications and location technologies, data processing and multisensor fusion and the paradigms for intelligent/adaptive management, which will make 'feasible' the development of advanced applications for ambient intelligence. On one hand, the concept of 'feasibility' gathers together technology and service requirements and business model viability. On the other hand, the broad concept of 'ambient intelligence' should be

here understood as one where mobile services, whichever they are, are provided to
the final user in a transparent, ubiquitous, continuous and personalized way.

From the conceptual point of view, this programme intends to contribute to the
'user-centric' approach associated to ambient intelligence, from the 'network-
centric' approach of cooperation and resource-sharing on which distributed systems
are based ('shared situational awareness', networks of users and ad-hoc networks).

From the application point of view, the programme intends to contribute to the
development of context-aware services in security, crisis management and disaster
recovery (civil defence, security forces, etc.), air traffic and airports (enhanced
surface mobility, intelligent management), tourism and natural environment, tele-
medicine and telecare (hospitals and smart homes) and e-mobility applications
(large scope applications). Technologies developed in the CONTEXTS project
will be validated in these challenging scenarios, which offer wide possibilities to
deliver advanced applications based on new interaction and fusion technologies.

José M. Molina   Ana M. Bernardos
José R. Casar   Antonio Berlanga
Felipe Cátedra
Javier Ortega-García
CONTEXTS'11 Program Co-chairs   CONTEXTS'11 Organizing Co-chairs

Organization

General Co-chairs

José M. Molina (Chairman)	University Carlos III of Madrid, Spain
José Ramón Casar Corredera	Polytechnic University of Madrid, Spain
Manuel Felipe Cátedra Pérez	University of Alcalá, Spain
Javier Ortega-García	Autonomous University of Madrid, Spain

Advisory Board

George Cybenko	Dartmouth College, USA
Antonio Ortega	Univ. Of Southern California, USA
James Llinas	State Univ. Of New York at Buffalo, USA
Ana C. Bicharra	Federal Fluminense University, Brasil
Raj Mitra	Pennsylvannia State Univ. USA
Stephano Maci	Univ. of Siena, Italy
Davide Maltoni	Univ. Bologna, Italy
Anil, K Jain	Michigan State Univ., USA
Gregori Vázquez	Pol. Univ. of Cataluña, Spain
Luis Vergara	Pol. Univ. of Valencia, Spain
Luis Correia	University of Lisboa, Portugal

Program Committee

Álvaro Luis Bustamante	University Carlos III of Madrid, Spain
Ana M. Bernardos	Polytechnic University of Madrid, Spain
Ana Cristina Bicharra	University Federal Fluminense, Brasil
Antonio Berlanga	University Carlos III of Madrid, Spain
Antonio Ortega	University of Southern California, USA
Carlos Delgado	University of Alcalá, Spain
Changjiu Zhou	Singapore Polytechnic, Singapore
Daniel Ramos Castro	Autonomous University of Madrid, Spain
David Griol	University Carlos III of Madrid, Spain

Doroteo Torre Toledano	Autonomous University of Madrid, Spain
Eleni Mangina	University College Dublin, Ireland
Eliseo García	University of Alcalá, Spain
Enrique Martí Muñoz	University Carlos III of Madrid, Spain
George Cybenko	Dartmouth College, USA
Gonzalo de Miguel	Polytechnic University of Madrid, Spain
Gonzalo Blazquez Gil	University Carlos III of Madrid, España
Gregori Vázquez	Polytechnic University of Cataluña, Spain
James Llinas	State University of N.Y. at Buffalo, USA
Javier Carbó	University Carlos III of Madrid, Spain
Javier Galbally Herrero	Autonomous University of Madrid, Spain
Javier Portillo	Polytechnic University of Madrid, Spain
Jesús García-Herrero	University Carlos III of Madrid, España
Joaquín González Rodríguez	Autonomous University of Madrid, Spain
Jørgen Bach Andersen	Aalborg University, Denmark
José Luis Guerrero	University Carlos III of Madrid, Spain
Jose Manuel Gómez	University of Alcalá, Spain
Juan Besada	Polytechnic University of Madrid, Spain
Juan M. Corchado	University of Salamanca, Spain
Juan Pavón	Complutense University of Madrid, Spain
Julián Fiérrez Aguilar	Autonomous University of Madrid, Spain
Juan Gómez Romero	University Carlos III of Madrid, Spain
Luis Correia	Lisboa University, Portugal
Luis Vergara	Polytechnic University of Valencia, Spain
Miguel Ángel Patricio	University Carlos III of Madrid, Spain
Miguel Serrano Mateos	University Carlos III of Madrid, Spain
Nayat Sánchez	University Carlos III of Madrid, Spain
Oscar Gutiérrez	University of Alcalá, Spain
Paula Tarrío	Polytechnic University of Madrid, Spain
Raj Mitra	Pennsylvania State University, USA
Rodrigo Cilla Ugarte	University Carlos III of Madrid, Spain
Rubén Vera Fernández	Autonomous University of Madrid, Spain
Virginia Fuentes	University Carlos III of Madrid, Spain

Organizing Committee

Ana M. Bernardos	Polytechnic University of Madrid, Spain
Antonio Berlanga	University Carlos III of Madrid, Spain

Contents

Integrating Multicamera Surveillance Systems into Multiagent Location Systems

José Luis Guerrero, Antonio Berlanga, and José M. Molina

Abstract. Users are increasingly demanding personalized services based on their context, being one of the key features of that context the user's position. There are a wide number of possible solutions to deal with the positioning issue, which, for different situations, may have different accuracy requirements. This paper presents this issue from the point of view of an existing multicamera surveillance system which requires to be integrated into a multiagent positioning system, including a tracking example with the presented architecture.

Keywords: Multicamera systems, PTZ cameras, multiagent systems, tracking.

1 Introduction

A well known trend in nowadays computing is the requirement of users to obtain different ranges of services according to their position, preferences, past actions, etc, which is usually known as context aware computing. Pioneer works in this field in the early nineties [1], [2] introduced and used concepts like ubiquous or pervasive computing [3], dealing with the automatic availability of different computers in an invisible way to the user.

The definition of context can be divided into different categories, regarding the information processed. In [4] the three main categories considered were computing context (accessible services and communication issues), user context (user location and profile) and physical context (external conditions). In [5] an additional important category is included, the time context (current date and time). The definition included at the beginning of this introduction deals with the most sensible category for the final user, its own context.

Among the user context responsibilities, one of the key processes is to be able to successfully determine the user location. Location systems [6] have been designed

José Luis Guerrero · Antonio Berlanga · José M. Molina
Group of Applied Artificial Intelligence (GIAA), Computer Science Department
University Carlos III of Madrid
Colmenarejo, Spain
e-mail: {joseluis.guerrero, antonio.berlanga,
josemanuel.molina}@uc3m.es

J.M. Molina et al. (Eds.): User-Centric Technologies and Applications, AISC 94, pp. 1–9.
springerlink.com   © Springer-Verlag Berlin Heidelberg 2011

for this task, both for indoor and outdoor locations. Probably the best known location system for outdoor positioning is GPS [7], whereas indoor location systems may range from radio frequency locating systems [8] to active bats approaches [9]. An important feature to compare the different available systems is the infrastructure requirements which the different locating systems require in order to perform their location procedures. In surveillance scenarios the most common infrastructure found is a set of cameras (currently commonly used for human guided surveillance) leading to the possibility of positioning using that already installed set of cameras in order to automate and improve the surveillance task [10].

Along with the positioning system, context aware systems may require a tracking procedure to determine the user's current position, according to the current information provided by the sensing device and the previous positions obtained. This need leads to the use of methods which can handle the inaccuracies of the positioning system and predict the user position according to certain mathematical models. Kalman filter [11] is one of the most extended techniques used for this task, even though (at least in its basic version) it is only suited for linear movement tracking.

As previously overviewed, different locating systems may provide the system with different information sources with different accuracies and different characteristics. When choosing a single one of those systems does not provide us the location quality required by our system, we may resort to a joint use of different systems, trying to keep their different benefits while minimizing their handicaps. This is performed by information fusion systems [12], which combine the information of these different systems in a variety of ways depending on the requirements of the final locating system. These systems frequently use some tracking function as an intermediate step in their fusion cycle.

The operation of the context aware system will require different services (which may or may not have dependencies among them) to be run at the same time, which, along with the ubiquous computing statements, requires to set up a distributed computing architecture. For this particular task, multi-agent systems [13] are particularly well designed, since they allow an easier automatic adaptation to the different environment situations which systems may develop. In [14], the benefits of developing multi agent systems as an information fusion system to guide Unmanned Aerial Vehicles (UAV) [15] are discussed, while in [16] a multi agent based system for location is presented.

The objective of this paper is to present the required architecture and functions to include multi-camera based surveillance system into a context-aware architecture. This architecture will deal with the control requirements for the different cameras, the access interfaces, both locally and remotely, and finally present an example regarding a tracking system for an object based on its color print.

The structure of the paper will be divided in the following sections: initially a system overview will be presented, detailing the components of the built system and detailing the architecture presented. The general architecture will be followed by a section containing the detailed proposal for the automated handling of the different cameras, leading to a final example showing the overall function of the presented system and the conclusions which the previous sections lead to.

2 System Overview

The concrete system used may add or remove certain requirements regarding the inclusion of a surveillance based system into a multi agent context aware system. This section will detail the concrete components used in the built system, along with their introduced restrictions and the general architecture design.

The cameras used are a set of Pan-Tilt-Zoom (PTZ) cameras which are Sony's VISCA protocol [17] compliant, such as the one showed in figure 1. The concrete models used where not able to provide the system with digital captures of their images, so digitalizer cards where used for this process, in particular Matrox Morthis frame grabbers. This introduces a handicap in the system, since the image provided by those cards may be required both by processes being run in the camera control and by agents external to it, and these cards can provide the image only to a single process. To enable a good scalability of the system regarding the images handling, a provider agent for each of the cards is built, being the images obtained by this intermediate provider, and thus preventing the direct access to the card's library.

Fig. 1 VISCA protocol compliant PTZ camera

The access to the camera's library was distributed into three different levels, regarding the processes performed at each level: the first level basically transcribes the function information to sets of bytes and sends it through the connected port. The second level provides an easy interface to access those low level functions, while the third is responsible of higher level control functions (such as zone limits and management).

The access to the controlling library may be performed on a local or remote way. To allow this, we have included a server wrapper which allows any process to send the required command and receive its response remotely, while, at the same time, the server itself operates as any other local process in its calls to the library. This implies the definition of a communication protocol between the server and the clients. The defined protocol is a simple alternative consisting in the command performed, the result (three digits) and the possibly required data (according to different commands). The different parts of this message are separated with the proper tokens to allow the parsing in the client. The message structure is overviewed in figure 2.

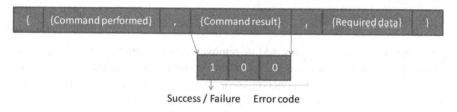

Fig. 2 Remote communication protocol

An important feature is that the calls performed to the library are, by definition of the protocol, blocking, which means that no additional command may be performed by the camera until its previous one has been completed. The architecture presented is summarized in figure 3

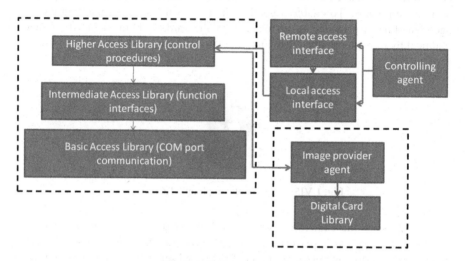

Fig. 3 Architecture overview

3 PTZ Cameras Control Library

Building a library for the control of the different cameras implies both general and particular requirements, which at the same time may be closer to the actual functions which the camera may perform directly or require more intermediate computations apart from the actual device. To account for these different characteristics, the designed library has been divided into three different levels: basic, intermediate and higher access levels.

The basic level offers to the intermediate one the functions which the device can perform directly (such as moving to a specific position), converts them to its equivalent byte string, sends them over the port, receives the response from the device and converts it to a set of possible responses which are interpreted by the intermediate level library. The actual packets sent may have a length varying from three to sixteen bytes, with the structure shown in figure 4. The actual implementation of this basic library level has used the open source library libvisca[1].

The intermediate library level offers friendly interfaces to the low level functions, handling of the responses from the device and the required actions according to them and some basic device specific control. An example of this control may be to determine the real movement bounds of the controlled device (something which cannot be performed by the basic library). If an order to move beyond the camera's boundaries in either of its axis is commanded, this intermediate level will detect it, move the camera as far as possible and response its caller with the proper information (indicating the wrong movement command and the action taken).

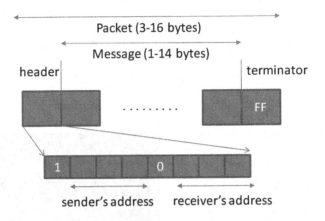

Fig. 4 VISCA packet format

The higher access library introduces some not strictly device related actions into the library, allowing useful functions such as zone processing. Zone processing makes the system able to switch to previously established zones without requiring the caller to remember their coordinates. This also helps to develop surveillance routines moving through different zones of interest by a set of waypoints. These procedures may be very domain specific, according to the purpose of the system in which the library will be included, so they have been implemented at a different level to promote the reusability of the previous levels not including possibly unnecessary code. Another example of these higher routines may be the one performed by the agent in the example included in the following section.

4   A Color Based Object Tracker

Previous sections presented the architecture design and library details of the proposed system. In this section we will apply the explained to the task of following an object according to its color print. The color print of an object is the color

---

[1] Available online at http://damien.douxchamps.net/libvisca/

components which the object has when it enters into our system. This approach can be useful in environments where the background of the image is prone to change (such as non-static cameras as the ones used in this paper). Depending on the application, it can be based on pre-established color prints or register the object color print and determine whether it is relevant or not once it enters the vision range of the given camera.

This function, even though here is presented as an agent procedure, can be considered as a higher level function of the camera library, since it can be useful in a multi agent surveillance system to coordinate the coverage of different cameras while one of them is performing the tracking of a certain object of interest.

The first level of this function performs the required image processing [18], obtaining the image to analyze and determining whether the interest color print is present in the image. The open computer vision library (OpenCV) [19] has been used to simplify this task. A color based filtering is performed, according to the histogram obtained from the color print, and then a linear filtering is performed according to the zone of interest's position, obtaining both the expected position of the object and the zone where the function musk look for it in the next iteration. Figure 5 shows an example of the color print filtering performed.

Fig. 5 Example of the color print image filtering

The next action involves determining the amount of movement required by the camera in order to follow the object. Alternatives based on fuzzy logic [20] were tested, allowing a smooth control over the camera movement when the object was static, but they did not achieve good results tracking movement over a certain speed (probably due to the introduced delay which their calculations required). The final implemented solution performs a simpler and faster approach based only on the boundaries of the object, its center and the center of the image (moving the camera the required amount so that the boundaries of the object would include the center of the image).

The linear filtering is suitable for the task due to its low complexity, but it leads to what is usually called *"disengagement"*: sudden movement changes make the filter predicted position wrong and thus it cannot find the object in its delimited zone of interest. To deal with this problem, a response based on two steps has

Fig. 6 Example of a tracking situation. From the top image on the left, the object of interest starts to be tracker, on the following one it has been centered, followed by a zoom on it when its size is reduced and a final zoom out when it is lost

been implemented: once the object of interest has not been found in the delimited zone, the whole image is searched for it, resetting the tracker state. If it cannot be found in the whole image, the camera resets its zoom status and goes back to its normal procedure (which might be to start a surveillance through different zones or stay in that position until another object matching the introduced color print is found).

5 Conclusions

Reusing available resources is always a complex situation, especially when that reuse involves the inclusion of an automation process. Even so, this is the common trend for many currently human controlled systems. In this paper we have highlighted the versatility requirements for a surveillance system in order to integrate it into a multiagent positioning system, proposed an architecture based on three different library levels to cope with those requirements and shown an example of the results by means of a color print tracking system. Future lines cover the analysis of the whole positioning system performance, including quantitative measurements of the benefits of including the different information sources, in particular the ones provided by the integrated multicamera system.

Acknowledgments. This work was supported in part by Projects CICYT TIN2008-06742-C02-02/TSI, CICYT TEC2008-06732-C02-02/TEC, CAM CONTEXTS (S2009/TIC-1485) and DPS2008-07029-C02-02.

References

1. Want, R., Hopper, A., Falcão, V., Gibbons, J.: The Active Badge location system. ACM Transactions on Information Systems 10(1), 91–102 (1992)
2. Schilit, B., Theimer, B., Welch, B.: Customizing mobile applications. In: Proceedings of USENIX Mobile & Location-Independent Computing Symposium, pp. 129–138. USENIX Association (August 1993)
3. Weiser, M.: The computer for the 21st century. Scientific American, 94–104 (September 1991)
4. Schilit, B., Adams, N., Want, R.: Context-aware computing applications. In: Proceedings of IEEE Workshop on Mobile Computing Systems and Applications, pp. 85–90. IEEE Computer Society Press, Santa Cruz (1994)
5. Chen, G., Kotz, D.: A Survey of Context-Aware Mobile Computing Research. Dartmouth Computer Science Technical Report TR2000-381 (2000)
6. Hightower, J., Borriello, G.: Location systems for ubiquitous computing. IEEE Computer Journal 34(8), 57–66 (2002)
7. Groves, P.D.: Principles of GNSS, Inertial, and Multisensor Integrated Navigation Systems. Artech House, Boston (2008)
8. Bahl, P., Padmanabhan, V.N.: RADAR: An in-building RF-based user location and tracking system. In: Proceedings of the Nineteenth Annual Joint Conference of the IEEE Computer and Communications Societies (2002)
9. Harter, A., Hopper, A., Steggles, P., Ward, A., Webster, P.: The Anatomy of a Context-Aware Application. In: Proceedings of the Fifth Annual International Conference on Mobile Computing and Networking, pp. 59–68 (1999)
10. López de Ipiña, D., Mendonça, P., Hopper, A.: TRIP: A Low-Cost Vision-Based Location System for Ubiquitous Computing. Personal and Ubiquitous Computing 6(3), 206–219 (2002)
11. Kalman, R.E.: A new approach to linear filtering and prediction problems. Transactions of the ASME- Journal of Basic Engineering (82), 35–45 (1960)
12. Waltz, E., Llinas, J.: Multisensor data fusion. Artech House, Boston (1990)
13. Koski, A., Juhola, M., Meriste, M.: Syntactic Recognition of ECG Signals by Attributed Finite Automata. Pattern Recognition 28(12), 1927–1940 (1995)
14. Guerrero, J.L., García, J., Molina, J.M.: Multi-agent Data Fusion Architecture Proposal for Obtaining an Integrated Navigated Solution on UAV's. In: Proceedings of the tenth International Work-Conference on Artificial Neural Networks: Part II: Distributed Computing, Artificial Intelligence, Bioinformatics, Soft Computing, and Ambient Assisted Living, pp. 13–20 (2009)
15. Valavanis, K.P.: Advances in Unmanned Aerial Vehicles. State of the Art and the Road to Autonomy. International Series on Intelligent Systems, Control and Automation: Science and Engineering, vol. 33. Springer, Heidelberg (2007)

16. Luis, A., Molina, J.M., Patricio, M.A.: Multi-Camera and Multi-Modal Sensor Fusion, an Architecture Overview. In: de Leon, F., de Carvalho, A.P., Rodríguez-González, S., De Paz Santana, J.F., Rodríguez, J.M.C. (eds.) Distributed Computing and Artificial Intelligence. Advances in Intelligent and Soft Computing, vol. 79, pp. 301–308. Springer, Heidelberg (2010)
17. Sony Corporation. Sony EVI-D70/D70P Technical Manual (2003)
18. Fisher, R., Dawson-Howe, K., Fitzgibbon, A., Robertson, C., Trucco, E.: Dictionary of Computer Vision and Image Processing. John Wiley, New York (2005)
19. Bradski, G., Kaehler, A.: Learning OpenCV: Computer Vision with the OpenCV Library. O'Reilly Media, Sebastopol (2008)
20. Klir, G.J., Yuan, B.: Fuzzy sets and fuzzy logic: theory and applications. Prentice Hall, Englewood Cliffs (1995)

Evaluating Manifold Learning Methods and Discriminative Sequence Classifiers in View-Invariant Action Recognition

Rodrigo Cilla, Miguel A. Patricio, Antonio Berlanga, and José M. Molina

Abstract. This paper evaluates the accuracy of Isometric Projections and Hidden Conditional Random Fields in the view invariant Recognition of Human Actions. Silhouette sequences captured from different viewpoints are projected into a low dimensional manifold using Isometric Projections. The projected sequences are used to train a hidden conditional random field for action classification. The system is evaluated using sequences captured by a camera not used during training. The accuracy of the system is measured using the IXMAS dataset on the experiments.

1 Introduction

Human Action Recognition from video has been one of the most important research topics studied by the computer vision community during the last two decades. Video Surveillance, Ambient Intelligence, Multimedia Annotation or Human Computer Interaction are some of the applications that have been benefited from the advances in the field. Different surveys recently published give an idea of the current state of the art methods [7, 21, 14]. These surveys also identify some of the issues that must be solved in order to obtain general human action recognition systems. How to achieve viewpoint invariant systems is one of them, as scene viewpoint is one of the greatest factors of variation of the visual features used to detect human actions. Different actions could even appear to be the same if observed from the wrong viewpoint.

The main benefit of a view independent human action recognition system is that it allows the transferability of models between cameras: it can use images grabbed from a camera whose viewpoint was not used during training without degrading its performance. This makes the deployment of the systems easier, as they not have to be retrained for the new views.

Rodrigo Cilla · Miguel A. Patricio · Antonio Berlanga · José M. Molina
Computer Science Deparment, Universidad Carlos III de Madrid.
Avda. de la Universidad Carlos III, 22. 28270 Colmenarejo, Madrid, Spain
e-mail: {rcilla,mpatrici}@inf.uc3m.es,
{berlanga,molina}@ia.uc3m.es

J.M. Molina et al. (Eds.): User-Centric Technologies and Applications, AISC 94, pp. 11–18.
springerlink.com

Graph Based Manifold learning methods[19, 1] have arisen as a powerful tool for the analysis and visualization of high dimensional data. They have been used as part of human action recognition systems [17, 8]. The main drawback of these techniques is that they do not provide a projection function from the original space to the reduced space. To overcome that limitation linear extensions of the methods have been proposed [3, 5].

Hidden Conditional Random Fields (HCRF) [15] have been introduced as a sequence classification technique, outperforming Hidden Markov Models (HMM). The main advantage of this models is that directly model the posterior distribution of the sequence class, using less parameters than HMM, allowing faster and more accurate parameter estimation. Its use on the recognition of Human Action has been very popular since its introduction[25, 11].

In this paper we evaluate the accuracy of graph based manifold learning and HCRF in the view invariant recognition of human actions. Human silhouettes are projected into a low dimensional space learned using Isometric Projections [3]. HCRF are trained using the projected sequences. The accuracy of the system is evaluated using action sequences coming from a new camera not used during system training.

Paper is organized as follows: in section 2 we discuss the state of the art on view-invariant action recognition; the graph based manifold learning framework is presented in section 3; section 4 presents the Hidden Conditional Random Field for sequence classification; section 5 presents the results of applying the proposed methods for the classification of the IXMAS dataset: finally, in section 6, the conclusions of this work and future research lines are presented.

2 Related Work

The problem of viewpoint action recognition has been studied from the geometrical perspective. Rao et al. [16] introduce a 2D view- invariant descriptor for 3D point trajectories projected in the affine plane. They search the spatio-temporal trajectory curvature to find instants of change. Parameswaran and Chellappa [12] present 2D and 3D invariants for body pose configurations. Gritai et al. [4] propose a metric to compare the trajectories of body parts under anthropometric, temporal and viewpoint transforms. Sheikh et al. [18] approximate the variability in action data as a linear combination of different action bases in spatio-temporal space. The main drawback of these approaches is that they assume that an accurate 3D tracking of the body parts is available, and this is very difficult to achieve in a real scenario.

The 3D visual hull offers an inherent view-invariant representation of human actions and has been used on multiple works. Weinland et al. [24] extend Motion History Images to 3D, using visual hulls instead of silhouettes to create Motion History Volumes. Peng et al. [13] perform multilinear analysis of visual hull voxels. Turaga et al. [22] analyzes visual hulls in Stieffel and Grassman manifolds. The main problem associated with visual hulls is that its computation requires from multiple camera views of the analyzed subject. Some authors have tried to relax this assumption for the prediction of new action sequences, associating observed silhouettes with the corresponding hidden visual hull. Lv and Nevatia [9] propose Action Nets, modeling in a space state the variation on silhouette appearance and viewpoint with respect to temporal evolution. Weinland et al. [23] propose a graphical model to associate observed silhouettes with the respective hidden visual hulls, incorporating camera parameters into the model.

Other authors have relied on machine learning techniques to output view- invariant action models. Martinez-Contreras et al.[10] project motion history images (MHI) [2] into a subspace that groups viewpoint and movement in a principal manifold using Kohonen self-organizing feature maps. The winner neuron is used to classify the action being performed using HMM smoothing. Tran et al.[20] proposed another approach to achieve view invariance, where view-invariant models are learned using non-parametric classification from a frame descriptor extracted from multiple views including appearance and local motion information.

Graph based manifold learning techniques have been used to learn view-invariant models. Souvenir and Babbs [17] propose a new silhouette based temporal action descriptor and embed them into a low dimensional space using Isomap [19]. The main drawback of their method is the assumption that target action sequences are already segmented. Lewandowski et al. [8] use temporal laplacian eigenmaps to obtain an embedding for each view. Then, the different embeddings are combined into a single one. While all these methods use visual hulls or camera configuration at some point, the proposed here only uses silhouette sequences. Besides, the proposed method provides a mapping function as output, without needing an additional method to compute it.

3 Isometric Projection

Given a set of N input points $X = \{x_i\}_{i=1}^{N}$, $x_i \in R^n$, Isometric projection (IsoP) [3] aims to find a function f that maps these N points to a set of points $Y = \{y_i\}_{i=1}^{N}$ in R^d ($d << n$), such $y_i = f(x_i)$. The method is of special applicability when the points $X \in M$, where M is a nonlinear manifold embedded in R^n.

Let d_M be the geodesic distance measure on M and d the standard Euclidean distance measure in R_d. IsoP aims to find an embedding function f such that Euclidean distances in R_d can provide a good approximation to the geodesic distances on M. Thus, the function to obtain is the one that minimizes:

$$\sum_{ij} (d_M(x_i, x_j) - d(f(x_i), d(x_j)))^2 \tag{1}$$

As the underlying manifold M where real dataset are defined is unknown, the geometric distance measure d_M is also unknown. To discover the intrinsic geometrical structure of M, a neighborhood graph G containing all the points in X is constructed. There is two standard ways to construct this graph:

1. ε-graph: A link is established between x_i and x_j if $d(x_i, x_j) < \varepsilon$
2. kNN-graph: A link is established between each point x_i and its k nearest neighbors.

Once the neighborhood graph G has been obtained, the geodesic distance on the manifold d_M between two points x_i, x_j is approximated as the distance on the graph between the points, $d_G(x_i, x_j)$. Thus, the matrix $D_G = d_G(x_i, x_j)$, containing the distances between all the points on X needs to be computed. The standard procedure to obtain D_G is to use Floyd-Warshall algorithm, with complexity $O(N^3)$. As the graph G is usually sparse, Johnson's algorithm can reduce the problem to $O(NV \log V)$, where V is the number of edges in G.

If the projection function f is restricted to be a linear function of the form $f(x) = A^T x$, the columns of the matrix $A = [a_1 \ldots a_d]$ are given by the solutions of the generalized eigenvalue problem:

$$X[\tau(D_G)]X^T a = \lambda X X^T a \tag{2}$$

where the matrix $\tau(D_G) = -HSH/2$ is an inner product matrix, being $H = I - \frac{1}{N} ee^T$, I the identity matrix and e a vector of all ones; and S a matrix such $S_{ij} = D_{ij}^2$.

Readers are referred to the original publication [3] for additional derivation and implementation details.

4 Hidden Conditional Random Fields

Hidden Conditional Random Fields (HCRF) [15] extend Conditional Random Fields [6] introducing hidden state variables into the model. A HCRF is an undirected graphical model composed of three different set of nodes, as figure 1 shows. The node y represents the labelling of the input sequencer. $X = x_1, \ldots, x_t$ is the set of nodes corresponding to the sequence observations $H = h_1, \ldots, h_t$ is the set of hidden variables modelling the relationship between the observations x_i and the class label y and the temporal evolution of the sequence.

The conditional probability of a sequence label y and a set of hidden part assignments $\mathbf{h}$ given a sequence of observations X is defined using the Hammersley-Clifford theorem of Markov Random Fields:

$$P(y, \mathbf{h} \mid \mathbf{x}, \theta) = \frac{e^{\Psi(y, \mathbf{h}, \mathbf{x}; \theta)}}{\sum_{y'} \sum_h e^{\Psi(y, \mathbf{h}, \mathbf{x}; \theta)}} \tag{3}$$

where θ is the vector of model parameters. The conditional probability of the class label y given the observation sequence X is obtained marginalizing over all the possible assignments of hidden parts $\mathbf{h}$:

$$P(y \mid \mathbf{x}, \theta) = \frac{\sum_h e^{\Psi(y, \mathbf{h}, \mathbf{x}; \theta)}}{\sum_{y'} \sum_h e^{\Psi(y', \mathbf{h}, \mathbf{x}; \theta)}} \tag{4}$$

The potential function $\Psi(y, \mathbf{h}, \mathbf{x}; \theta)$ is a linear function of the input:

$$\Psi(y, \mathbf{h}, \mathbf{x}; \theta) = \sum_i \phi(x_i) \cdot \theta(h_i) + \sum_i \theta(y, h_i)$$
$$+ \sum_{(j,k) \in E} \theta(y, h_j, h_k) \tag{5}$$

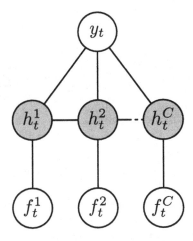

Fig. 1 Graphical model representation of the Hidden Conditional Random Field

The first term, parameterised by $\theta(h_i)$ meassures the compatiblity of each observation x_i with the hidden variable h_i. The second term measures the compatiblity of the hidden part h_i with the class label and is parameterised by $\theta((y,h_i))$. Finally, the third term models sequence dynamics, meassuring the compatibility of adjacent hidden parts h_i and h_j with the class y.

Given a set of training samples $\{x^i, y^i\}$, model parameters are adjusted maximizing the L_2 regularized conditional likelihood function of the model:

$$L(\theta) = \sum_{i=1}^{n} \log P(y_i \mid \mathbf{x}_i, \theta) + \frac{||\theta||^2}{2\sigma} \qquad (6)$$

The optimal parameters θ^* maximizing the conditional likelihood function are found using Quasi-Newton gradient based methods. Both the computation of the posterior probability on equation 4 and the auxiliary distributions that appear on the gradient of 6 can be efficiently made using belief propagation, as proposed in [15].

5 Results

5.1 Experimental Setup

The proposed algorithms are going to be tested in the classification of IXMAS dataset [24]. This dataset contains 11 actions performed by 12 different actors at least 3 times each. The actions are recorded from 5 different viewpoints. In our experiments we are going to discard the camera 5 viewpoint, as it shows a top view of the action completely different from the others. The algorithms are going to be trained using Leave-One-View-Out Cross Validation (LOVO-CV): The algorithms are trained with all the views unless one, used for validation. With this method the robustness of classification algorithm is evaluated with respect to changes on the viewpoint.

(a) Camera 1   (b) Camera 2   (c) Camera 3   (d) Camera 4

The silhouettes on the dataset are resized to fit a bounding box of 35x20 pixels. They are going to be projected into 10, 20 and 30 dimensions using Isometric Projections. The neighbourhood graph is going to be built using a neighborhood size of $k = 20$. PCA projections of the same size are going to be used as baseline. The HCRF is going to be trained with 22 and 33 hidden state variables.

5.2 Results and Discussion

Table 1 shows the accuracy of the trained classifiers. IsoP achieves better results than PCA for most of the configurations. It is worth pointing that the best accuracy is obtained by IsoP for the smaller dimension configuration, being a bit superior to the best found for PCA, but using one third of the dimensions. We also want to point that the use of more hidden nodes in the HCRF has increased the predictive power of the final result.

Table 1 Results obtained in the LOVO-CV evaluation of the proposed methods on IXMAS dataset

$\|H\|$ $\diagdown$ d	PCA			IsoP		
	10	20	30	10	20	30
22	39.42	49.26	46.47	49.81	53.34	52.41
33	48.76	53.77	55.88	**56.00**	55.94	53.03

In any case, the results presented here are inferior to those reported on other works [23]. IsoP algorithm imposes a restriction on the projection function to be linear, and the kind of transform need to achieve view invariance is likely not to be. Probably the use of other algorithms providing non linear projections will improve the results.

6 Conclusions

In this work we have evaluated the use of Isometric Projections and Hidden Conditional Random Fields in the view invariant recognition of human actions.

Human silhouette sequences have been projected into a low dimensional space using Isometric Projections. Then, a Hidden Conditional Random Field has been trained using that sequences. The accuracy of the system has been obtained using action sequence taken from viewpoints not used during training.

Future works will have to test other manifold learning algorithms, as the predictive power of the proposed system has not been very high. Kernelized versions of the proposed algorithms can be one way to explore.

Acknowledgment

This work was supported in part by Projects CICYT TIN2008-06742-C02-02/TSI, CICYT TEC2008-06732-C02-02/TEC, CAM CONTEXTS (S2009/ TIC-1485) and DPS2008-07029-C02-02.

References

1. Belkin, M., Niyogi, P.: Laplacian eigenmaps for dimensionality reduction and data representation. Neural computation 15(6), 1373–1396 (2003)
2. Bobick, A.F., Davis, J.W.: The recognition of human movement using temporal templates. IEEE Transactions on Pattern Analysis and Machine Intelligence 23(3), 257–267 (2001)
3. Cai, D., He, X., Han, J.: Isometric projection. In: Proceedings of the National Conference on Artificial Intelligence, vol. 22, p. 528. AAAI Press, MIT Press, Menlo Park, CA, Cambridge, MA, London (1999)
4. Gritai, A., Sheikh, Y., Shah, M.: On the use of anthropometry in the invariant analysis of human actions. In: Proceedings of the 17th International Conference on Pattern Recognition (ICPR 2004), vol. 2, pp. 923–926 (2004)
5. He, X., Niyogi, P.: Locality Preserving Projections. In: NIPS (2003)
6. Lafferty, J., McCallum, A., Pereira, F.: Conditional random fields: Probabilistic models for segmenting and labeling sequence data. In: International Conference on Machine Learning (2001)
7. Lavee, G., Rivlin, E., Rudzsky, M.: Understanding Video Events: A Survey of Methods for Automatic Interpretation of Semantic Occurrences in Video. IEEE Transactions on Systems, Man and Cybernetics - Part C: Applications and Reviews 39(5), 489–504 (2009)
8. Lewandowski, M., Makris, D., Nebel, J.C.: View and Style-Independent Action Manifolds for Human Activity Recognition. In: Daniilidis, K., Maragos, P., Paragios, N. (eds.) ECCV 2010. LNCS, vol. 6316, pp. 547–560. Springer, Heidelberg (2010)
9. Lv, F., Nevatia, R.: Single view human action recognition using key pose matching and viterbi path searching. In: IEEE Conference on Computer Vision and Pattern Recognition, 2007 (CVPR 2007), pp. 1–8. IEEE, Los Alamitos (2007)
10. Martínez-Contreras, F., Orrite-Uruñuela, C., Herrero-Jaraba, E., Ragheb, H., Velastin, S.A.: Recognizing Human Actions using Silhouette-based HMM. In: IEEE Conference on Advanced Video and Signal-based Surveillance, pp. 43–48 (2009)
11. Mendoza, M., Pérez De La Blanca, N.: Applying space state models in human action recognition: a comparative study. Articulated Motion and Deformable Objects, 53–62 (2008)

12. Parameswaran, V., Chellappa, R.: View invariants for human action recognition. In: Proceedings IEEE Computer Society Conference on Computer Vision and Pattern Recognition, 2003, vol. 2 (2003)
13. Peng, B., Qian, G., Rajko, S.: View-Invariant Full-Body Gesture Recognition via Multilinear Analysis of Voxel Data. In: Third ACM/IEEE Conference on Distributed Smart Cameras (September,2009)
14. Poppe, R.: A survey on vision-based human action recognition. Image and Vision Computing 28(6), 976–990 (2010)
15. Quattoni, A., Wang, S., Morency, L.-P., Collins, M., Darrell, T.: Hidden conditional random fields. IEEE Transactions on Pattern Analysis and Machine Intelligence 29(10), 1848–1852 (2007)
16. Rao, C., Yilmaz, A., Shah, M.: View-invariant representation and recognition of actions. International Journal of Computer Vision 50(2), 203–226 (2002)
17. Richard, S., Kyle, P.: Viewpoint Manifolds for Action Recognition. EURASIP Journal on Image and Video Processing (2009)
18. Sheikh, Y., Sheikh, M., Shah, M.: Exploring the space of a human action. In: Tenth IEEE International Conference on Computer Vision, 2005 (ICCV 2005), vol. 1 (2005)
19. Tenenbaum, J.B., Silva, V., Langford, J.C.: A global geometric framework for nonlinear dimensionality reduction. Science 290(5500), 2319 (2000)
20. Tran, D., Sorokin, A., Forsyth, D.: Human activity recognition with metric learning. In: Forsyth, D., Torr, P., Zisserman, A. (eds.) ECCV 2008, Part I. LNCS, vol. 5302, pp. 548–561. Springer, Heidelberg (2008)
21. Turaga, P., Chellappa, R., Subrahmanian, V.S., Udrea, O.: Machine Recognition of Human Activities: A Survey. IEEE Transactions on Circuits and Systems for Video Technology 18(11), 1473–1488 (2008)
22. Turaga, P., Veeraraghavan, A., Chellappa, R.: Statistical analysis on Stiefel and Grassmann manifolds with applications in computer vision. In: IEEE Conference on Computer Vision and Pattern Recognition (CVPR 2008), pp. 1–8. IEEE, Los Alamitos (2008)
23. Weinland, D., Boyer, E., Ronfard, R.: Action recognition from arbitrary views using 3d exemplars. In: IEEE 11th International Conference on Computer Vision, 2007(ICCV 2007), pp. 1–7. IEEE, Los Alamitos (2007)
24. Weinland, D., Ronfard, R., Boyer, E.: Free viewpoint action recognition using motion history volumes. Computer Vision and Image Understanding 104(2-3), 249–257 (2006)
25. Zhang, J., Gong, S.: Action categorization with modified hidden conditional random field. Pattern Recognition 43(1), 197–203 (2010)

Physical Actions Architecture: Context-Aware Activity Recognition in Mobile Devices

Gonzalo Blázquez Gil, Antonio Berlanga,
and José M. Molina

Abstract. Mobile phones are becoming the main computer and communication device in peoples lives and thanks to the embedded sensors providing in it will revolutionize the way to carry out with mobile devices. Another important point is that mobile devices are programmable and they provide a set of embedded sensors, such as accelerometer, digital compass, gyroscope, GPS, microphone, and camera. Activity recognition aims to recognize actions and goals of one or more individual from a series of observations of themselves. This paper aims to provide an distributed architecture to recognize physical actions taken by users such us walking, running, being stand, sitting. Sensory data is collected by a mobile application made in Android and it sends to a server where prelearnt activities are recognized in real-time.

1 Introduction

Daily, human beings make ordinary actions such us: cooking, reading or watching TV, chatting with other people or on the phone , driving. The ability of activity recognition seems so natural and simple for us, however, actually requires complicated functions of sensing, learning, and inference for computers [7].

In activity recognition applications, high classification accuracy is always desired. However, it implies the use of a large number of sensors distributed over the body and enviroment, depending on the activities to detect. At the same time a wearable system must be inconspicuous and operate during long periods of time. This implies minimizing sensor size, and especially low energy consumption.

Gonzalo Blázquez Gil · Antonio Berlanga · José M. Molina
Applied Artificial Intelligence Group, Universidad Carlos III de Madrid,
Avd. de la Universidad Carlos III, 22, 28270, Colmenarejo, Madrid, Spain
e-mail: {gonzalo.blazquez,antonio.berlanga,josemanuel.molina}@uc3m.es
http://www.giaa.inf.uc3m.es

J.M. Molina et al. (Eds.): User-Centric Technologies and Applications, AISC 94, pp. 19–27.
springerlink.com

Pattern recognition answer to the description and classification of measurements taken from physical or mental processes [9] [11]. In order to provide an effective and efficient description of patterns, preprocessing is often required improve performance, removing noise and redundancy in measurements. Then a set of characteristic measurements, which could be numerical or not, and relations between them, are extracted representing the patterns.

Thanks to the newest mobile devices use this kind of sensor could be a good solution to face with activity recognition problem [2]. Mobile devices may obtain and process physical phenomena from embedded sensors and send sensory data to remote locations without any human intervention. GPS, Wi-Fi, Bluetooth and microphone are the most known sensors. However, recently, new kind sensors have been added: accelerometer, gyroscope, compass (magnetometer), proximity sensor, light sensor, etc.

Sensor networks have contributed to numerous attractive applications in areas such as military applications, environmental monitoring, human activity, medical applications, ambient assistant living, smart factories, civil security [8] [10] [1]. Applications for a mobile devices should take advantage of contextual information, such as position, user profile or device features; to offer greater services to the user.

In [5], it is presented a mobile sensing architecture to obtain user physical activities and to share it on social network called Cenceme. The proposed architecture is split in three layer: Sense, learn and share. Sense layer aims to collect raw sensor data from sensors embedded in the phone. However, Cenceme does not use other information (Context information) to complete the taken actions.

Context-aware Systems allow to develop a new kind of mobile applications which may be represented as a context based scenario where there are individuals who require a satisfaction of their needs and there are providers who could solve these lacks. In context-aware computing, context is any information that can be used to characterize the situation of an entity.

Context-aware in general describes applications, which change their behavior according to the conditions around them. Positioning, also called Location context, could be the main factor in the development of context applications. Nevertheless, Location-aware is only one aspect of context aware as a whole. Profile Context which include social networking, Calendar scheduling, tastes, moods; and device context which represents the actions taken by the user carrying the device, are becoming essential to describe the user's context.

According to the three described context (Profile, Location and Device), it is defined three kind of actions (Social, Location and Physical). Initially, this paper is concerned with one part of user's context, device context and Physical actions. The actions to take into account are walking, running, standing, sitting (using Accelerometer and compass) and audio actions such us listening to music and talking.

This paper is focused on describe an mobile-server architecture to identify physical actions (e.g., running, walking, standing, talking, ...) taken by the user. Besides, it is developed the feature selection for physical actions. This paper is organized as follows: Section 2 describes how to obtain the context from different sensors. Section 3 presents the proposal architecture and finally, Section 4 shows the conclusions and futures works.

2   Context-Aware Description

First of all, in order to use context correctly, it is mainly to define what researchers think context is. According to Lee [6] there are three kind of context (Figure 1):

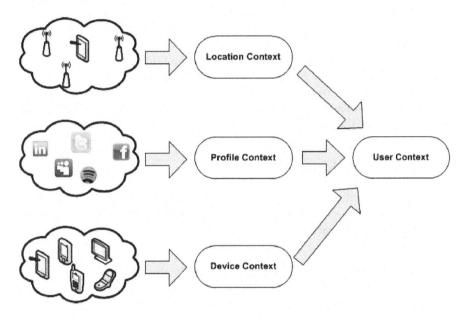

Fig. 1 Three sort of context (Profile, Device, Location).

1. Location context, which can be described as an application dependent on a geographical location. Probably, location context is the most important factor to develop useless Context-aware applications. In outdoor environments GPS provides a good solution to determine the location of mobile devices however in GPS-denied areas such as urban, indoor, and subterranean environments, unfortunately, an effective solution does not exist. Besides, every location system provides in its own way location data. Recently, W3C releases a Geolocation API [4] to standardize an interface to get back the geographical location information for a client device.
2. User context is the relevant information about the user, normally, it is included in the user profile as an instance. Social networks play important roles in our daily lives. People share their own information through

social networks with relatives. It is possible to leverage social media such as Facebook, Buzz, Twitter, and Flickr as ways to not only disseminate information but obtain user activity information.

3. Device context shows the need to adapt services and content to different devices due to their limitations on screen size, memory, wireless connection and the mobile devices movements. For example, The proximity and light sensors allow the phone to perform context recognition associated with the display.

- User behavior: Active applications, installed application, idle/active status, phone alarm, charger status.
- User surroundings: In this case it is talked about physical surrounding, Bluetooth/Wi-Fi devices online near user or sociologic surroundings such as friends, family, colleagues, collaborators, and business partners.
- Communication behavior: Calls and call attempts, sent and received SMS, and SMS content, social network communication.
- Device actions: Due to embedded sensors attached to the mobile device it is possible to obtain the movement. The information generated can be used to identify the activity (e.g., running, walking, standing, and so on) that the user is performing.

3 General Architecture

According to the three kind of context described in the last section (Profile, Device and location Context), it is possible to distinguish three actions (Social, Physical and Location actions), each action is represented by a kind of context. (Figure 2):

- Physical actions are the basic actions taken by the user such us: running, walking, standing, talking, listening music, etc. These kind of actions are obtained by low level sensors provided by the mobile phone (Accelerometer, Gyroscope, light sensor, microphone, etc.). For example, accelerometer is able to describe the physical movements of the user carrying the phone.
- Location actions: These kind of actions answer the question where is the action taking place?. For example, it is possible to define a running action, however, it could be interested to define where is he/she running? and where is he/she running to?.
- Social actions: Finally Social actions describe relationships between people and also it describes the user preferences Thanks to these actions it could be possible to complete Context activity describing the people with you are taking the action.

Context activity is the result of combining the three actions. As well as user context, it describes the general action, it is not only if the user is walking or he/she is in one place. For example, consider the following scenario, someone is sitting in her/his living-room watching TV.

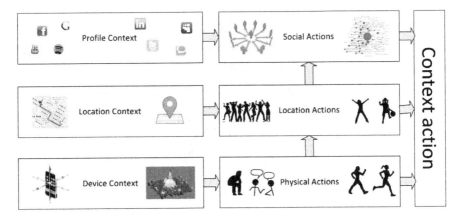

Fig. 2 Matching between context and actions.

The accelerometer and the microphone may detect whether the user is sitting (Physical action) or the user is near a sound source (Physical action). If you use the both actions and it is able to locate the action (living room) it could figure out that the person is sitting in the living room watching TV (Location action).

3.1 Pyshical Actions Architecture

Mobile phone sensing is still in its infancy, it is not clear what architectural components should run on the device and what should run on the cloud. This approximation split the architecture in three component according to main components of activity recognition (Figure 3).

Data Acquisition and Features extraction are developed in the mobile phone and Activity Recognition is located on the server. Due to the communication process between them it is developed a Web service in the server and another component in the mobile phone to send the extracted features to the web service.

3.1.1 Data Acquisition

A low-level sensing module continuously gathers relevant information about the user activities using sensors. The proposed architecture this component is located on the mobile device. In this approximation the architecture just acquire data from Compass and Accelerometer. The accelerometer provides the forces (static and non-static) acting on the device. It returns a three component (x ,y ,z) vector that represent the three-axis forces acting on the device Cartesian reference system. Note that the accelerometer reference system is also constantly changing due to device's position.

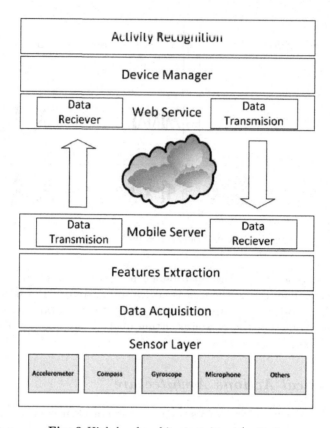

Fig. 3 High-level architecture input/output.

However, it is necessary to obtain that forces according to the real world, for that reason it is used the compass which provide us the I matrix (Inclination matrix). I matrix transform the coordinates from the device reference to the real world reference.

The first graph (Figure 4) represents the device accelerations and shows change the three forces depending on the movement take it by the user. The second one represents the transformation from the mobile device reference to the real world reference.

3.1.2 Features Extraction

The features extraction level is also implemented in the mobile phone. The module processes the raw sensor data into features that help discriminate between activities. This level aims to process and select which features are better to identified an action.

It is used the spectrogram (Figure 5) to define Physical actions. A spectrogram is a a time-varying spectral representation that shows how the spectral

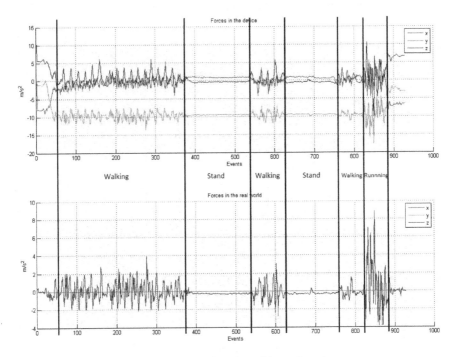

Fig. 4 Device and real world accelerations.

density of a signal varies with time. Spectrograms was calculated from the time signal (Figure 4) using the short-time Fourier transform (STFT).

The active frequencies (Red and yellow bars) depending on the action took it are clearly different between running and walking actions. Non active frequencies are colored in blue and normally it represents when the frequencies are not active. If the whole spectrogram is blue represent when users is doing a sedentary action (Standing, sitting, etc).

3.1.3 Mobile Server, Web Service and Device Manager

The both components aims to communicate the mobile device with the server. One of both(Mobile server) is implemented in the mobile devices and the other one (Web service) is on the server.

The Web service module is developed such as web service which is designed to support interoperable machine-to-machine communication over a network. Web services provide an interface which describe message the format, specifically, Web Services Description Language WSDL [3].

Device manager allows web-service to view and control the devices attached to the service. When a device is not online, the web-server keep the last device's IP address for a while, waiting for a new connection.

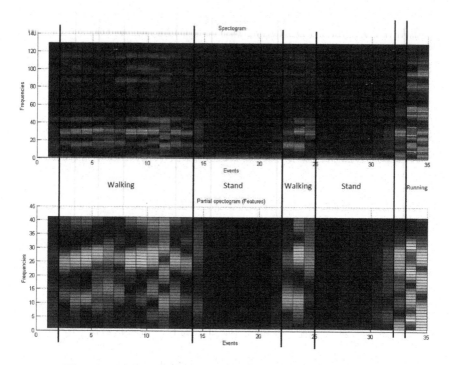

Fig. 5 Complete and parcial spectogram (Features selection).

3.1.4   Activity Recognition

The last layer is classification module that uses the features selected in the mobile phone to infer what activity an individual or group of individuals is engaged in, for example: Walking, running, sitting, standing. In this component, it will be implemented the algorithms (Supervised learning, Probabilistic classification, Model-based or instance-based learning) to figure out the taken action.

4   Conclusions

The activity recognition system identifies and record in real-time the selected features related on user activity using a mobile device. The paper describes how to face with the sensing module, that is one of the major components in activity recognition systems.

Considered future works include the development of the server module and also it will extend activity classifier to more complex activities (Group activities, Interaction activities). Using Context activity (Physical, Social and Location) will be use to infer the user's emotional state. To infer human emotional state from user context and the taken activities is a novel line of research. According to the social network state, the music which is listening at the moment, the place where the user is and using another hard and low sensor, it is possible to infer the emotional state.

Acknowledgments

This work was supported in part by Projects CICYT TIN2008-06742-C02-02/TSI, CICYT TEC2008-06732-C02-02/TEC, CAM CONTEXTS (S2009/TIC-1485) and DPS2008-07029-C02-02.

References

1. Consolvo, S., McDonald, D., Toscos, T., Chen, M., Froehlich, J., Harrison, B., Klasnja, P., LaMarca, A., LeGrand, L., Libby, R. et al.: Activity sensing in the wild: a field trial of ubifit garden. In: Proceeding of the twenty-sixth annual SIGCHI conference on Human factors in computing systems, pp. 1797–1806. ACM Press, New York (2008)
2. Eagle, N., Pentland, A.: Reality mining: sensing complex social systems. Personal and Ubiquitous Computing 10(4), 255–268 (2006)
3. World Wide Web Consortium Editors. Web services description language 1.1 @ONLINE (March 2001)
4. World Wide Web Consortium Editors. Geolocation api specification @ONLINE (February 2010)
5. Lane, N., Miluzzo, E., Lu, H., Peebles, D., Choudhury, T., Campbell, A.: A survey of mobile phone sensing. IEEE Communications Magazine 48(9), 140–150 (2010)
6. Lee, W.: Deploying personalized mobile services in an agent-based environment. Expert Systems with Applications 32(4), 1194–1207 (2007)
7. Liao, L.: Location-based activity recognition. PhD thesis, Citeseer (2006)
8. Mun, M., Reddy, S., Shilton, K., Yau, N., Burke, J., Estrin, D., Hansen, M., Howard, E., West, R., Boda, P.: Peir, the personal environmental impact report, as a platform for participatory sensing systems research. In: Proceedings of the 7th international conference on Mobile systems, applications, and services, pp. 55–68. ACM Press, New York (2009)
9. Simon, J.: Recent progress to formal approach of pattern recognition and scene analysis. Pattern recognition 7(3), 117–124 (1975)
10. Thiagarajan, A., Ravindranath, L., LaCurts, K., Madden, S., Balakrishnan, H., Toledo, S., Eriksson, J.: VTrack: accurate, energy-aware road traffic delay estimation using mobile phones. In: Proceedings of the 7th ACM Conference on Embedded Networked Sensor Systems, pp. 85–98. ACM, New York (2009)
11. Verhagen, C.: Some general remarks about pattern recognition; its definition; its relation with other disciplines; a literature survey. Pattern recognition 7(3), 109–116 (1975)

Context-Aware Conversational Agents Using POMDPs and Agenda-Based Simulation*

David Griol and José M. Molina

Abstract. Context-aware systems in combination with mobile devices offer new opportunities in the areas of knowledge representation, natural language processing and intelligent information retrieval. Our vision is that natural spoken conversation with these devices can eventually become the preferred mode for managing their services by means of conversational agents. In this paper, we describe the application of POMDPs and agenda-based user simulation to learn optimal dialog policies for the dialog manager in a conversational agent. We have applied this approach to develop a statistical dialog manager for a conversational agent which acts as a voice logbook to collect home monitored data from patients suffering from diabetes.

1 Introduction

Ambient Intelligence (AmI) systems usually consist of a set of interconnected computing and sensing devices which surround the user pervasively in his environment and are invisible to him, providing a service that is dynamically adapted to the interaction context. In this framework, spoken interaction can be the only way to access information in some cases, like for example when the screen is too small to display information (e.g. hand-held devices) or when the eyes of the user are busy in other tasks (e.g. driving). It is also useful for remote control of devices and robots, specially in smart environments. Finally, one of the most demanding applications for fully natural and understandable dialogs, are embodied conversational agents and companions. This way, conversational agents have became a strong alternative to provide computers with intelligent and natural communicative capabilities.

David Griol · José M. Molina
Group of Applied Artificial Intelligence (GIAA). Computer Science Department. Carlos III University of Madrid
e-mail: {david.griol,josemanuel.molina}@uc3m.es

* Funded by projects CICYT TIN2008-06742-C02-02/TSI, CICYT TEC2008-06732-C02-02/TEC, CAM CONTEXTS (S2009/TIC-1485), and DPS2008-07029-C02-02.

J.M. Molina et al. (Eds.): User-Centric Technologies and Applications, AISC 94, pp. 29–36.

A conversational agent is a software that accepts natural language as input and generates natural language as output, engaging in a conversation with the user. To successfully manage the interaction with the users, conversational agents usually carry out five main tasks: automatic speech recognition (ASR), natural language understanding (NLU), dialog management (DM), natural language generation (NLG) and text-to-speech synthesis (TTS).

Learning statistical approaches to model the different modules that compose a conversational agent has been of growing interest during the last decade [7, 2]. The motivations for automating dialog learning are focused on the time-consuming process that hand-crafted design involves and the ever-increasing problem of dialog complexity. The most extended methodology for machine-learning of dialog strategies consists of modeling human-computer interaction as an optimization problem using Partially Observable Markov Decision Process (POMDPs) and reinforcement methods [5]. The main drawback of this approach is due to the large state space of practical spoken dialog systems, whose representation is intractable if represented directly. An approach that scales the POMDP framework for implementing practical spoken dialog systems by the definition of two state spaces and the use of an agent is presented in [6].

In this paper, we apply this approach to develop a conversational agent which acts as a voice logbook to collect home monitored data from patients suffering from Type-2 diabetes [1], integrated in a context-aware multiagent architecture [3]. In this architecture, a Context Manager in the conversational agents deals with loading the context information provided by the User and Positioning Agents, and communicates it to the different modules of the Conversational Agent during the interaction. Context information is managed by means of a data structure called *user profile*. The conversational agent developed using the POMDP and agenda-based user simulation has been compared with a dialog management technique based on a rule-based model. The results of this comparison show how the POMDP methodology achieves satisfactory task completion rates and improves the selection process of goal directed actions.

2   The DI@L-Log System

DI@L-log is a spoken dialog system which acts as a voice logbook to collect home monitored data from patients suffering from Type-2 diabetes [1]. The data collected by the system are the patient's weight, blood pressure (systolic and diastolic values) and sugar levels. The system validates and analyses the data, providing some immediate feedback to the patients regarding their current progress as well as communicating the results to doctors at the hospital who are able to review the patient's progress graphically and deal with any alerts generated by the system concerning abnormal developments.

To apply our methodology to this task, the definition of the semantics was carried out considering the information that is required to monitor the patients and inform them about their condition. Taking into account the requirements of the task, an

---

S1: Welcome to the DI@L-log System. Please say or enter your four digit pin.
U1: 1356.
S2: Are you ready to give your measurements?
U2: Yes.
S3: Let's begin with your weight. How many stones?
U3: 12.
S4: I have understood 12 stones. Is it correct?
U4: Yes.
S5: What is your blood sugar?
U5: 11.
S6: I have understood 11. Is it correct?
U6: Yes.
S7: And finally, for your blood pressure you will be asked for the top number and then the lower number. What is the top number of your blood pressure?
U7: 160.
S8: I have understood 160. Is it correct?
U8: Yes.
S9: And what is the lower number?
U9: 86.
S10: I have understood 86. Is it correct?
U10: Yes.
S11: Your readings have been successfully updated. According to them, your weight is higher than last week, be very careful. Thanks for contacting DI@L-log. Goodbye.

Fig. 1 An example of a dialog for the DI@L-log task

initial strategy was designed for the DM. Figure 1 shows an example of a dialog acquired using this strategy.

As can be observed, three different phases are present in every dialog. Firstly, there is an identification phase in which the system asks the user about his login and password and then waits until the user says that he is ready to provide the control data (S1 and S2 system turns). Secondly, the system analyzes which data is required for the current user, taking into account that the weight and sugar values are mandatory and the blood control is only carried out for specific patients (S3 to S10 system turns). In this phase, the system requires the user to provide this data. Every item is confirmed after the user has provided its value. The user can only provide one item at a time. In the last phase, the system consults the information that the patient has provided during the current dialog and compares it with the data that is present in a database that contains the values that he provided in previous dialogs. By means of this comparison, the system is able to inform the user about his condition and provide him with instructions that take this into account (S11 system turn).

A corpus of 100 dialogs was acquired using this strategy. In order to learn statistical models, the dialogs of the corpus were labeled in terms of dialog acts. In the case of user turns, the dialog acts correspond to the classical frame representation of the meaning of the utterance. For the DI@L-log task, we defined three task-independent concepts (*Affirmation*, *Negation*, and *Not-Understood*) and four attributes (*Weight*, *Sugar*, *Systolic-Pressure*, and *Diastolic-Pressure*).

The labeling of the system turns is similar to the labeling defined for the user turns. A total of 12 task-dependent concepts was defined, corresponding to the set of concepts used by the system to acquire each of the user variables (*Weight*,

Sugar, *Systolic-Pressure*, and *Diastolic-Pressure*), concepts used to confirm the values provided by the user (*Confirmation-Weight*, *Confirmation-Sugar*, *Confirmation-Systolic*, and *Confirmation-Diastolic*), concepts used to inform the patient about his condition (*Inform*), and three task-independent concepts (*Not-Understood*, *Opening*, and *Closing*).

3   POMDPs and Agenda-Based User Simulation

Formally, a POMDP is defined as a tuple $\{S, A, T, R, O, Z, \lambda, b_0\}$ where:

- S is a set of the agent states.
- A is a set of actions that the agent may take.
- T defines the transition probability $P(s'|s, a)$.
- R defines the immediate reward obtained from taking a particular action in a particular state $r(s, a)$.
- O is a set of possible observations that the agent can receive.
- Z defines the probability of a particular observation given the state and machine action $P(o'|s', a)$.
- λ is a geometric discount factor $0 \leq \lambda \leq 1$.
- b_0 is an initial belief state $b_0(s)$.

The operation of a POMDP is as follows. In each moment, the agent is in an unobserved state s. The agent selects an action a_m, receives a reward r, and transits to a state (unobserved) s', where s' only depends on s and a_m. The agent receives an observation o' which depends on s and a_m. Although the observation allows the agent to have some evidences about the state s in which the agent is now, s is not exactly known, and $b(s)$ (*belief state*) is defined to indicate the probability of the agent being in the state s. In each moment, this probability is updated taking into account o' and a_m:

$$b'(s') = P(s'|o', a_m, b) = k \cdot P(o'|s', a_m) \sum_{s \in S} P(s'|a_m, s)b(s) \qquad (1)$$

where $k = P(o'|a_m, b)$ is a normalization constant [4].

At each time t the agent receives a reward $r(b_t, a_{m,t})$ which depends on b_t and the selected action $a_{m,t}$. The reward accumulated during the dialog (*return*) can be calculated by means of:

$$R = \sum_{t=0}^{\infty} \lambda^t R(b_t, a_{m,t}) = \sum_{t=0}^{\infty} \lambda^t \sum_{s \in S} b_t(s)r(s, a_{m,t}) \qquad (2)$$

Each action $a_{m,t}$ is determined by the policy $\pi(b_t)$ and the construction of the POMDP model implies to find the strategy π^* which maximizes *return*. The goal of POMDP policy optimization is to find the policy that maximizes the value function at every point b. Due to the vast space of possible belief states, however, the use of POMDPs for any practical system is far from straightforward. Exact algorithms for solving POMDPs do exist, but have been shown to be intractable except for domains limited to a few states. Instead, the belief state and actions are mapped down to a summarized form where optimization becomes tractable. In this context, the original belief space and actions are called master space and master actions, while the summarized versions are called summary space and summary actions. The summarized space consists of the N-best states (s_u) from the original space (N is usually 1 or 2) and a simplified codification of the user's action a_u and the dialog history s_d. The main idea of this summarized space is to explore only the actions that has sense for the current situation in the dialog (e.g. do not begin the conversation with a confirmation, do not say *Welcome* except at the start, etc.).

The optimization of the policy in these two spaces is usually carried out by using techniques like the Point-based Value Iteration or Q-learning, in combination with a user simulator. Q-learning is a technique for online learning traditionally used in an MDP framework. It is an iterative Monte-Carlo style algorithm where a sequence of sample dialogs are used to estimate the Q functions for each state and action. This way, the summarized Q-learning algorithm discretizes summary space and uses Q-learning on the resulting MDP-like grid. Using this algorithm, at each point, the master belief space is mapped down to the summary level as described and the nearest summary point in the grid is found and the optimal summary action given by that point is chosen. This optimal action for each point p is given by

$$\bar{a}_p = \underset{\bar{a}}{\operatorname{argmax}} \bar{Q}(a, p)$$

After a set of dialogs has been completed, the estimates of the Q-functions are updated with the new dialog scores. At the end of the dialog, the discounted future reward is known for each stage where a choice was taken (i.e., the Q-function evaluated at this grid point). A good estimate of the true Q-value is obtained if sufficient dialogs are done. User simulation is then introduced to reduce the too time-consuming and expensive task to obtain these dialogs with real users. Simulation is usually done at a semantic dialog act level to avoid having to reproduce the variety of user utterances at the word or acoustic levels. At the semantic level, at any time t, the user is in a state s_u, takes action a_u, transitions into the intermediate state s'_u, receives machine action a_m, and transitions into the next state s''_u.

Agenda-Based state representations, like the one described in [6], factors the user state into an agenda A and a goal G. The goal G consists of constraints C which specify the detailed venue of the dialog, and requests R which specify the desired pieces of information.

$$s_u = (A; G) \quad G = (C; R)$$

The user agenda A is a stack-like structure containing the pending user dialog acts that are needed to elicit the information specified in the goal. At the beginning of each dialog, a new goal G is randomly selected. Then, the goal constraints C are converted into user and system *inform acts* (a_u and a_m acts) and the requests R into *request acts*. A *bye* act is added at the bottom of the agenda to close the dialog once the goal has been fulfilled. The agenda is ordered according to priority, with A[N] denoting the top and A[1] denoting the bottom item. This way, $a_u[i]$ denotes the ith item in the user act a_u:

$$a_u[i] := A[N-n+i] \quad \forall i \in [1..n]; \ 1 \leq n \leq N$$

$$P(a_u|s_u) = P(a_u|A,G) = \delta(a_u, A[N-n+1..N])$$

By denoting A' to the agenda after selecting the act a_u, the first state transition depending on a_u and the second state transition based on a_m can be respectively expressed by means of the following equations [6]:

$$P(s_u'|a_u, s_u) = P(A', G'|A, G) = \delta(A', A[1..N'])\delta(G', G)$$

$$P(s_u''|a_m, s_u') = P(a_m|A', G'')P(G''|a_m, G')$$

The first probability in the latter equation denotes the agenda update model. By assuming that every dialog act a_m triggers one push-operation from the agenda, this probability can be expressed as follows:

$$P(a_m|A', G'') = \prod_{M}^{i=1} P(A''[N'+i]|a_m[i], G'') \ \delta(A''[1..N'], A'[1..N'])$$

The second probability denotes the goal update model. Assuming that R'' is conditionally independent of C' given C'', it can be expressed as follows:

$$G''|a_m, G') = P(R''|a_m, R', C'')P(C''|a_m, R', C')$$

where the first probability can be approximated as follows:

$$P(R''|a_m, R', C'') = \prod_{k} P(R''[k]|a_m, R'[k], \mathscr{M}(a_m, C''))$$

4   Evaluation

The summary Q-learning algorithm and agenda-based user simulation described in the previous section were used to develop a POMDP-based conversational agent for the Di@L-log task. To do this, we took into account the benefits of using standards like VoiceXML and also include a specific module for the statistical dialog model to avoid the effort of manually defining the dialog strategy. This module selects the following system response using the dialog policy obtained by means of the POMDP. By means of the incorporation of this module, developers only have to define a set of VXML files, each one including a system prompt and the associated grammar to capture users answers for it. This way, the statistical dialog manager automatically decides the following file (i.e. system prompt) that has to be selected.

A total of 25 dialogs were recorded from interactions of six users employing the initial dialog strategy defined for the DI@L-log system and the POMDP-based systems presented in this paper. Rewards in this system were given based on the task completion rate and the number of turns in the dialog. Using the definitions described in [6], the POMDP system was given 20 points for a successful dialog and 0 for an unsuccessful one, One point was subtracted for each dialog turn. We considered the following measures for the evaluation:

1. Dialog success rate (%success). This is the percentage of successfully completed tasks. In each scenario, the user has to obtain one or several items of information, and the dialog success depends on whether the system provides correct data (according to the aims of the scenario) or incorrect data to the user.
2. Average number of turns per dialog (nT).
3. Confirmation rate (%confirm). It was computed as the ratio between the number of explicit confirmations turns (nCT) and the number of turns in the dialog (nCT/nT).
4. Average number of corrected errors per dialog (nCE). This is the average of errors detected and corrected by the dialog manager. We have considered only those errors that modify the values of the attributes and that could cause the failure of the dialog.
5. Average number of uncorrected errors per dialog (nNCE). This is the average of errors not corrected by the dialog manager. Again, only errors that modify the values of the attributes are considered.
6. Error correction rate (%ECR). The percentage of corrected errors, computed as nCE/ (nCE + nNCE).

The results presented in Table 1 show that the initially defined rule-based conversational agent and the POMDP-based conversational agent could interact correctly with the users in most cases. The success rate in the POMDP system is reduced with regard to the initial rule-based system. This is due to the introduction in this system of confidence scores to indicate the cases for which a confirmation must be done. This means that it is possible for the system to assign a high level confidence to an incorrectly recognized value. For this reason, the error correction rate in this system is also slightly lower. The average number of required turns is reduced in the POMDP-based system from 10.4 to 7.0. This is again due to the initial rule-based strategy is based on confirming every item provided by the user in the following system turn. The POMDP-based system reduces the number of required dialog turns by reducing the number of confirmations, as it can be observed in the lower confirmation rate achieved for this system.

Table 1 Results of the evaluation of the rule-based and POMDP-based conversational agents

	%success	nT	%confirm	%ECR	nCE	nNCE
Rule-based Conversational Agent	97%	10.4	41%	92%	0.81	0.07
POMDP-based Conversational Agent	93%	7.0	28%	89%	0.86	0.11

5   Conclusions

Modeling human-computer interaction by means of POMDPs and reinforcement
methods are the most extended methodology for machine-learning of dialog strategies in conversational agents. Due to the main drawback of this approach is the large
state space of practical spoken dialog systems, whose representation is intractable
if represented directly, we have applied the proposal described in [6] to deal with
this and also introduce an agenda-based model user simulation technique to learn
the dialog model. The application of this approach to develop a conversational agent
which collects home monitored data from patients suffering from diabetes show how
it allows the dialog manager to tackle new situations and generate new coherent answers for the situations already present in the initial corpus. Due to the new learning
process, the dialog manager can now ask for the required information using different
orders, confirm these information items taking into account the confidence scores,
reduce the number of system turns for the different kinds of dialogs, automatically
detect different valid paths to achieve each of the required objectives, etc. As a future work, we want to compare this approach with other statistical and corpus-based
methodologies for dialog management.

References

1. Black, L., McTear, M.F., Black, N.D., Harper, R., Lemon, M.: Appraisal of a conversational artefact and its utility in remote patient monitoring. In: Proc. of the 18th IEEE
   Symposium CBMS 2005, Dublin, Ireland, pp. 506–508 (2005)
2. Griol, D., Hurtado, L., Segarra, E., Sanchis, E.: A Statistical Approach to Spoken Dialog
   Systems Design and Evaluation. Speech Communication 50(8-9), 666–682 (2008)
3. Griol, D., Sánchez-Pi, N., Carbó, J., Molina, J.: An Architecture to Provide Context-Aware
   Services by means of Conversational Agents. Advances in Intelligent and Soft Computing 79, 275–282 (2010)
4. Kaelbling, L., Littman, M., Cassandra, A.: Planning and acting in partially observable
   stochastic domains. Artificial Intelligence 101, 99–134 (1998)
5. Levin, E., Pieraccini, R., Eckert, W.: A stochastic model of human-machine interaction
   for learning dialog strategies. IEEE Transactions on Speech and Audio Processing 8(1),
   11–23 (2000)
6. Thomson, B., Schatzmann, J., Weilhammer, K., Ye, H., Young, S.: Training a real-world
   pomdp-based dialogue system. In: Proc. of NAACL-HLT 2007, pp. 9–16 (2007)
7. Young, S.: The Statistical Approach to the Design of Spoken Dialogue Systems. Tech.
   rep., CUED/F-INFENG/TR.433, Cambridge University Engineering Department, Cambridge, UK (2002)

Multi-camera Control and Video Transmission Architecture for Distributed Systems

Alvaro Luis Bustamante, José M. Molina, and Miguel A. Patricio

Abstract. The increasing number of autonomous systems monitoring and controlling visual sensor networks, make it necessary an homogeneous (device-independent), flexible (accessible from various places), and efficient (real-time) access to all their underlying video devices. This paper describes an architecture for camera control and video transmission in a distributed system like existing in a cooperative multi-agent video surveillance scenario. The proposed system enables the access to a limited-access resource (video sensors) in an easy, transparent and efficient way both for local and remote processes. It is particularly suitable for Pan-Tilt-Zoom (PTZ) cameras in which a remote control is essential.

Keywords: multi-camera systems, visual sensor network, video transmission, ptz cameras.

1 Introduction

At the moment, the majority of the people still conceives video surveillance systems as synonymous of CCTV systems: people imagine tens of old cameras connected to tens of remote monitors, controlled by tens of bored and unheeding security employers which should pay attention to restricted areas, access doors, people, vehicles, objects and suspicious situations to prevent crimes or disasters. In alternative, many believe that surveillance systems are storage platforms to memorize multimedia data on the environment, video, photos, wiretapped speech, available for human forensic experts to support investigations. This is partially true, and the value of these systems is undoubted.

Alvaro Luis Bustamante · José M. Molina · Miguel A. Patricio
Applied Artificial Intelligence Group, Universidad Carlos III de Madrid,
Avd. de la Universidad Carlos III, 22, 28270, Colmenarejo, Madrid, Spain
e-mail: {alvaro.luis,miguelangel.patricio}@uc3m.es,
josemanuel.molina@uc3m.es
http://www.giaa.inf.uc3m.es

J.M. Molina et al. (Eds.): User-Centric Technologies and Applications, AISC 94, pp. 37–45.
springerlink.com   © Springer-Verlag Berlin Heidelberg 2011

However, latest advances in hardware technology and state of the art of computer vision and artificial intelligence research can be employed to develop autonomous and distributed monitoring systems. They are possible and necessary since the enormous improvement and afordability of hardware and the availability of distributed computing technologies have encouraged an increasing use of distributed and parallel systems in monitoring applications [4, 5]. So the growing amount of sensors sometimes makes unaffordable a human monitoring.

In addition, sensors are increasing their uses and capabilities with characteristics like on-board processing, Pan-Tilt-Zoom control, thermal and infrared vision, and so on [14]. It is valuable since it provides more features to the user, but it adds an extra control problem. For instance, the control of PTZ cameras can be easy to achieve in a small scale, but it can become a really tedious task when you have a large vision sensor network, which may lead in a poor usage of the available resources. A sample scenario may be one security employer attempting to monitor a specific moving target from several PTZ cameras simultaneously. The operator should reorientate the cameras in real-time according to the moving object, which result in a hard task to achieve.

This way, it would be useful that all these tedious operations were able to be managed by an autonomous system. It will help the operator who is working with the visual sensor network to exploit efficiently all the resources. Thereby a operator may set a new goal like *'find someone with a red bag'*, and the autonomous system should help in the different devices coordination, in order to meet the established goal.

Many researches has been focused in solve this and similar issues using multi-agent systems [11, 12], in where each agent is the responsible of control and manage one camera. This distributed solution is a good option for the problem of coordinating multi-camera systems, taking the advantages of scalability and fault-tolerance over centralization.

However, many of the proposed theoretical architectures uses to miss the underlying complex task of controlling the video sensors, like the image acquisition and transmission process [12, 8, 13]. They assume that in some way there is a video flow and a control flow for PTZ cameras which can be used in their architectures. However when dealing with a real implementation of a distributed multi-agent architecture with a real visual sensor network the problems arrives.

In order to support this kind of distributed systems, in this this paper is described the required architecture for manage the video devices present in a visual sensor network. This architecture will deal with the PTZ control and the video transmission for each video sensor connected, both for local and remote processes. This way, any camera will have their control and video accessible from any place. It is suitable for multi-agent systems since they really are distributed process and require remote control of the sensors. This kind of architecture involves many disciplines like video compression and transmission, advanced memory management, frame grabbers controllers, etc, thus, an overview and some test of the system designed is presented.

The rest of the paper is organized as follow. In section 2 we describes the environment in which the presented system may work. In section 3 is described the architecture supporting any distributed system. Section 4 contains some performance test, and finally in section 5 some conclusions are presented.

2 Environment Overview

Multimedia surveillance systems are an emerging application field requiring multidisciplinary expertise spanning from Signal and Image Processing to Communications and Computer Vision [4]. However, in this paper we focus on the problem of communication between the cameras and the different systems that control them, since cameras are limited-access resources.

Our initial working scenario consists in a visual sensor network in which each video device provides Pant-Tilt-Zoom (PTZ) control. It is intended to be used with a multi-agent system, so we want to provide an homogeneous access between different entities controlling this network.

In this scenario *a priori* may operate over the cameras two different entities. We define a operator which may be the personnel security monitoring the video streamed by different cameras and controlling their orientation. In a multi-agent architecture also exists the agents that generally perform an autonomous control depending on the restrictions/goals imposed by the operator.

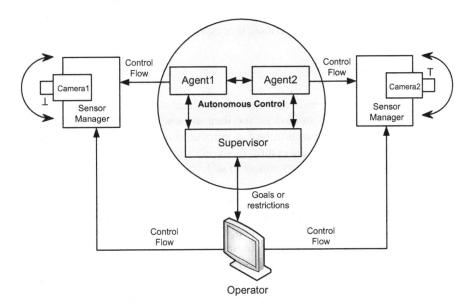

Fig. 1 Example of the environment overview. The camera devices are controlled both by an operator and a distributed system.

This kind of control over the cameras is shown in figure 1, where all the control flow presented in the architecture is described. It is created both from the operator and the multi-agent distributed system, so the devices must be able to handle requests from different sources.

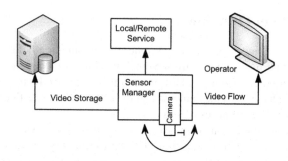

Fig. 2 Camera video transmission to different local and remote systems.

Figure 1 only presents the control flows, but in the architecture also exists video flows, since each camera device must be able to stream the video to different destinations. It may be necessary for remote monitoring, for storage platforms to memorize multimedia data on the environment, or simply for local or remote systems to perform advanced processing like video tracking [2], activity recognition [3], intrusion detection, etc. This is shown in figure 2. In this case all the flows out from the camera to the different local and remote processes.

3   Sensor Manager

In the previous section we have described the different roles of the architecture that will operate over the camera devices and their associated control/video flows. In the pictures describing the architecture is introduced a *Sensor Manager* which is the responsible of attending the control flows, allowing the positioning of the PTZ cameras, and stream the video sequences to all locally-attached process and remote systems.

Therefore the *Sensor Manager* is not a trivial component of the architecture, as almost all the multi-agent systems suppose. It involves many disciplines like video acquisition, compression and transmission; control the protocols of the PTZ device and expose it to remote controlling processes, and so on.

So, in this section is described the *Sensor Manager* designed for this task. We have taken special care to the real-time restriction of the video surveillance systems, in which the video stream should by delivered with the minimum delay. The overall architecture of this controller is presented in figure 3.

This design let multiple access to a limited-access PTZ camera device, what usually only provides one serial communication port for control the orientation of the camera, and a coaxial video interface with the analog video signal. The sensor

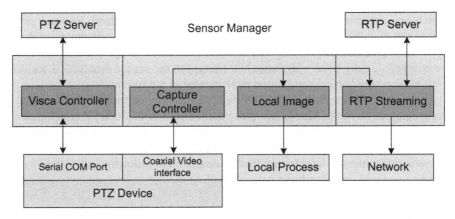

Fig. 3 Sensor manager for local/remote camera control

manager designed presents three different main functionalities which are described in more detail in the following subsections:

3.1 PTZ Controller

The PTZ controller designed allow us to provide an standard interface through the *PTZ Server* module to control homogeneously any underlying device. We have defined some high-level operating primitives like *goTo X Y*, *zoom amount*, etc, common in almost all PTZ devices. This commands will be interpreted by the proper controller (Visca Controller in the figure) and transmitted in the correct protocol to the device through the serial COM port.

High-level primitives are exposed by a non-connection oriented UDP Server, with a simple request-response protocol in the client-server computing mode. In our case we translate the high-level primitives incoming from the *PTZ Server* to the VISCA protocol [1], since all the cameras in our visual sensor network are compliant with this protocol. Moreover, any other device protocol may be used easily adapting a controller to our high-level interface.

3.2 Video Acquisition

Video acquisition from the device is a critical point of the sensor manager designed. In general, the image provided by present sensor devices comes in a analog format. This way is needed use digitizer cards to convert the signal to digital frames, in particular Matrox Morphis frame grabbers are used in our system. This introduces a handicap in the system, since the image provided by those cards is required by both local and external processes, while these cards can be handled only by a single process.

To solve this issue we have defined two different strategies depending on the image destination. For local processes we have shared a region of non-paged memory which can be accessed by different processes of the local machine. So local processes can read the latest acquired frame directly from memory without any delay. Other local processes that needs to be notified when a new image is available could be attached directly (with a dynamic link library) to the sensor manager, as we do with a color-based object recognition implemented in the system.

For remote process we use other efficient way to provide frames with the minimum delay. In this case, Matrox Morphis boards allows JPEG2000 [7] compression in real-time, so we can obtain a compressed image version at the same time we acquire uncompressed frames (used for local processes or display). These JPEG2000 compressed frames are later transmitted (as described in following subsection) by a efficient and real-time streaming protocol [10].

3.3   Video Transmission

In order to transmit real-time video sequences to remote process like operators, agents, backup systems, etc, we have opted to implement a Real-time Transport Protocol (RTP) server based on JPEG2000 image sequences, described in the newly RFC5371 [6], inside the sensor manager.

The RTP defines a standardized packet format for delivering audio and video over IP networks. RTP is used extensively in communication and entertainment systems that involve streaming media, such as telephony, video conference applications and in general all those applications what needs a real-time communication.

In the design of the sensor manager architecture a RTP payload header extension has been implemented following the RFC5371. This standard defines a new RTP extension allowing transmit JPEG2000 frames (provided by Matrox Morphis boards) over RTP packets. Moreover in our implementation of the standard we have introduced a real-time motion compensation technique which is still patent pending (Spanish patent number P200900260) and still complaints with RFC5371. Hence, our RTP server is able to transmit real-time JPEG2000 frames to many clients, both in a unicast and multicast way.

4   Architecture Evaluation

To evaluate the usability and performance of the architecture we have developed a functional prototype. It is divided in two different layers. The first layer is the sensor manager core, developed in different C++ modules (as described in the architecture) in order to meet real-time performance. The second one is the interface for configuring the core, as shown in figure 4. It provides both the RTP and PTZ servers, so any remote or local service may interact with the device concurrently. The presented Graphical User Interface (GUI) allows the user to select the capture device, camera channel, server ports, and some other parameters related with the JPEG2000 codec. It also enables a local PTZ controller for change the camera device orientation.

The sensor manager performance has been tested in two ways. In one hand we have measured the frames per second that the architecture is able to transmit both

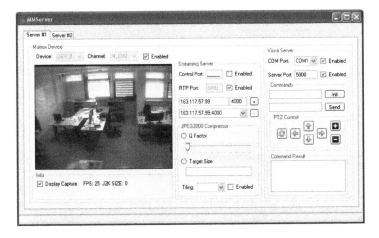

Fig. 4 Sensor Manager interface, providing controls over different aspects like RTP and PTZ servers

to local and remote processes. Each sensor manager controls one Matrox Morphis board (as the device only can be accessed by a single process at a time), and each board contains two hardware digitizers. It is possible to acquire video from 16 different sources, but in this case, it is necessary to share digitizers between all sources, hence the performance decreases. So we will use two sources for each board in order to find the real sensor manager performance, and not the limited by the underlying hardware.

Table 1 Transmission rate reached with the sensor manager architecture both for local and local/remote processes

Cameras	Local Process	Local/Remote Processes
One Camera	25 fps	25 fps
Two Cameras	25 fps	24-25 fps

Table 1 shows the different values obtained by the sensor manager in various operation modes, with one and two cameras transmitting video frames to local and remote processes. In both situations the frame-rate obtained is the same as the video sensor provides (25 FPS), so the architecture is working efficiently and in real-time when transmitting the video frames. If the architecture not were able to process all the frames or send it across the network, the digitizer board will start do drop frames and the frame-rate will significantly be reduced.

We have also measured the Matrox Morphis board usage in order to confirm that the architecture is not adding extra lag in process frames. Figure 5 shows the usage of one digitizer when sharing frames with local processes and when sharing/transmitting frames to local/remote processes. In both cases the grab unit is

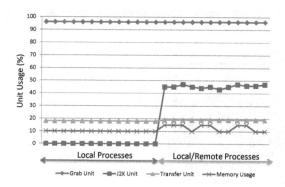

Fig. 5 Matrox Morphis unit usage under local and local/remote image transmission

always at 96%, meaning that all the video frames are being procsesed by the sensor manager. Notice also how the J2K unit start working only when is necessary transmitt frames to remote process.

In the other hand we have measured the total amount of time that the JPEG2000 streaming systems takes from when the image is captured in the server to when the image is displayed in the client. So it involves all the processes of acquisition, compression, transmission, decoding and displaying. The total latency obtained is about 180 milliseconds, which is adequate for real-time purposes like video surveillance, as discussed in [9].

Finally the PTZ server could not be evaluated since it process and execute all the high-level concurrent commands in sequence, so it depends on the commands executed, the camera movement speed, etc.

5 Conclusions

As outlined along this paper, the communication between local/remote processes with video sensor devices is so useful in the current growing visual sensor networks. It allows, among other things, to develop versatile architectures from cooperative multi-agent platforms, to sophisticated video surveillance systems. However the communication with a video device may be a unsatisfactory job since they used to have a limited-access and not always are accessible from remote places. An architecture for controlling video devices from local and remote processes in a transparent way has been proposed in this paper to solve this issue. The test performed show that the architecture is able to efficiently share video frames to several destinations.

In future works we will include some high-level commands in the PTZ control, like 'follow the red bag'. In this case, the image processing like color tracking will be addressed locally.

Acknowledgments

This work was supported in part by Projects CICYT TIN2008-06742-C02-02/TSI, CICYT TEC2008-06732-C02-02/TEC, SINPROB, CAM CONTEXTS S2009/TIC-1485 and DPS2008-07029-C02-02.

References

1. Sony EVI-D70/D70P Color Video Camera. Technical Manual
2. Black, J., Ellis, T., Rosin, P.: A novel method for video tracking performance evaluation. In: Joint IEEE Int. Workshop on Visual Surveillance and Performance Evaluation of Tracking and Surveillance (VS-PETS), Citeseer, pp. 125–132 (2003)
3. Cilla, R., Patricio, M., Berlanga, A., Molina, J.: Fusion of Single View Soft k-NN Classifiers for Multicamera Human Action Recognition. In: Corchado, E., Graña Romay, M., Manhaes Savio, A. (eds.) HAIS 2010. LNCS, vol. 6077, pp. 436–443. Springer, Heidelberg (2010)
4. Foresti, G., Mahonen, P., Regazzoni, C.: Multimedia video-based surveillance systems: Requirements, issues, and solutions. Springer, Netherlands (2000)
5. Foresti, G., Regazzoni, C., Varshney, P.: Multisensor surveillance systems: The fusion perspective. Kluwer Academic Publishers, Dordrecht (2003)
6. Futemma, S., Itakura, E., Leung, A.: RTP Payload Format for JPEG 2000 Video Streams. RFC 5371 (Informational) (October 2008)
7. ISO/IEC. 15444-1:2000 information technology jpeg2000 image coding system-part 1: core coding system. Technical report (2000)
8. Kang, S., Paik, J., Koschan, A., Abidi, B., Abidi, M.: Real-time video tracking using PTZ cameras. In: Proc. of SPIE, Citeseer, vol. 5132, p. 103
9. Karlsson, G.: Asynchronous transfer of video. IEEE Communications Magazine 34(8), 118–126 (2002)
10. Luis, A., Patricio, M.: Scalable Streaming of JPEG 2000 Live Video Using RTP over UDP. In: International Symposium on Distributed Computing and Artificial Intelligence 2008 (DCAI 2008), pp. 574–581. Springer, Heidelberg (2008)
11. Orwell, J., Massey, S., Remagnino, P., Greenhill, D., Jones, G.: A multi-agent framework for visual surveillance. In: Proceedings International Conference on Image Analysis and Processing 1999, pp. 1104–1107. IEEE, Los Alamitos (2002)
12. Patricio, M., Carbo, J., Perez, O., Garcia, J., Molina, J.: Multi-agent framework in visual sensor networks. EURASIP Journal on Applied Signal Processing 2007(1), 226 (2007)
13. Stillman, S., Tanawongsuwan, R., Essa, I.: A system for tracking and recognizing multiple people with multiple cameras. In: Proceedings of the Second International Conference on Audio-Vision-based Person Authentication, Citeseer (1999)
14. Wolf, W., Ozer, B., Lv, T.: Smart cameras as embedded systems. Computer 35(9), 48–53 (2002)

A Structured Representation to the Group Behavior Recognition Issue

Alberto Pozo, Jesús Gracía, Miguel A. Patricio, and José M. Molina

Abstract. The behavior recognition is one of the most prolific lines of research in recent decades in the field of computer vision. Within this field, the majority of researches have focused on the recognition of the activities carried out by a single individual, however this paper deals with the problem of recognizing the behavior of a group of individuals, in which relations between the component elements are of great importance. For this purpose it is exposed a new representation that concentrates all necessary information concerning relations peer to peer present in the group, and the semantics of the different groups formed by individuals and training (or structure) of each one of them. The work is presented with the dataset created in CVBASE06 dealing the European handball.

Keywords: Group behavior recognition, activity representation, computer vision.

1 Introduction

Human activity analysis and behavior recognition has received an enormous attention in the last two decades of computer vision community. A significant amount of research has addressed to behavior recognition of one element in the scene.

Instead of modeling the activities of one single element, group behavior recognition deals with multiple objects and/or people, who are part of groups.

Group behavior recognition represents a relatively new interesting direction of research, which has many possible applications in different situations like group sports, surveillance, defense, ethology, etc.

In behavior recognition there are two distinct philosophies for modeling a group; the group could be dealtas a single group (crowd) or as a composition of individuals with some shared objectives. In this paper we focus the investigation in the second philosophy, where take place many distinguishable agents.

Alberto Pozo · Jesús Gracía · Miguel A. Patricio · José M. Molina
GIAA, Carlos III University, Spain
e-mail: {alberto.pozo, jesus.garcia, miguelangel.patricio,
josemanuel.molina}@uc3m.es

J.M. Molina et al. (Eds.): User-Centric Technologies and Applications, AISC 94, pp. 47–57.

Individual behavior recognition and group behavior recognition have differences we should consider. Group behavior is not the addition of multiples individual's behaviors, instead of, group behavior depends on individual's activities, relations between the elements and roles played by each element.

The recognition of the dynamic of the groups can be applied in many complex areas, such as sports, security, ethology and defence. This recognition also can be extrapolated to any field of research, composed by autonomy agents which behaviour needs to be studied.

The present paper shows a new representation of the possible variables existed in the problem. This had been designed to put in order briefly the essential information of the system.

With the aim of achieve our project, it will rely on three levels of abstraction.

Firstly, a matrix will be established to store away the polar coordinates position at the system and the binary relationship existed between them. They will be represented by the free vectors which connect both individual.

In these terms and conditions, for each frame in the video, the geometrical information has being kept in two or three dimensions.

Once being contained all the geometrical information, the process continues in a second abstraction's level where the challenge is capturing the logical information implicated between the communication of individual and groups. For this reason it is necessary to make different combinations for representing every group of the system.

It is a relevant detail to remark that each individual can belong to a several groups at the same time, and the groups have the possibility to incorporate an undefined number of other groups or individual.

In the third level, a new representation is created to reduce the dimension of the problem. One of the important key in this type of domains is that the number of relations between the elements of the scene growth exponential in relation with the number of elements. For this reason, a new representation is created to save the formation information of each group without saving all the relations between each element.Instead of save all the possible edges in a graph, this approach only save important graph that can provide all the formation information wasting less space.

The paper is organized as follows. Section 2 reviews related work. Section 3 describes the problem. Section 4 introduces our description. Conclusions are drawn in section 5.

2 Related Work

Despite the fact thatthere is plenty of work on single object activities(1), (2), (3), the field of group activities recognition is relatively unexplored.

Group behavior recognition is a complex task that can be observed under different points of view.

There are two big families of approaches, one logical and one geometrical.

The logical approaches (4) are focused in construct a context-free grammar to describe the group activity based in the individual activities of the group's member.

The main characteristic of this point of view is the important of the first level, the features extraction. They need a previous system that recognizes the activity of each element of the scene.

The geometrical approaches (5), (6) have a different point of view. The features extracted in this case are based on the coordinates of the elements of the scene.

This approaches use to have higher computational complexity and the number of the elements in the scene could become very important.

There are also approaches than combine both perspective, like (7) whose work recognize key elements of a sport (basketball) using the positions of the players. This approach needs to identify the key elements of the domain dealt, and these key elements could be different in many different situations.

One more general approach could be read in (8) where the trajectories of the players (in a Robocup mach) are coded to create set of patterns that identify each type of action.

3 Group Behavior Recognition Issue: A General Overview

Group behavior recognition is composed by two steps: in the first one the features of the system should be extracted, and in the second one the features are used to recognize the behavior.

The system could have a lot of types of features like position, individual action, trajectory, speed, color, etc.

In this paper we are going to focus in the second step, we use the extracted features by other system to make the inference of the behavior.

3.1 General Description

In a general scene exist one area composed by many sub-areas and a number of elements (which could be fixed or not).

Each element of the system has a set of features (like positioning, color, shape, etc.) The element's features could suffer changes in time.

Each element of the system should belong to a group, and could belong to many groups at the same time.

It is important to emphasize that any element of the system must be in a group, so there are not isolated elements.

Each group has an internal and an external attitude. Each attitude could be cooperative or competitive. Internal attitude defines the attitude between the members of a group, and external attitude defines the attitude of the group respecting of the rest of the groups.

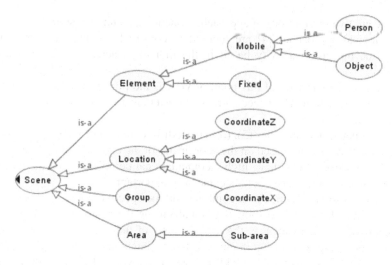

Fig. 1 General scene

Fig. 1 shows the general scene described above.

3.2 Problem Description

Some of the general axioms of the problem describe above have been eliminated for more practical approach of the problem.

In our approach there is one sequence of video composed by a number of T frames, where are included a number of N elements (this number cannot change in time). The elements of the scene are distributed in a number of G groups, and each group is represented by a graph. (Group could be composed by two or more elements, and one element could be part of one or more groups).

Each node constitute an agent of the group, and each graph is represented by his edges.

One edge is represented by free vector expressed in polar coordinate system.

Fig. 2 shows a scene with six elements conforming three groups. The definition of the groups is the semantic representation of the relations between the elements of the system.

The features selected to describe the elements of the scene are its coordinates (in polar coordinate system or spherical coordinate system in case of 3-D positioning) and the coordinates of the free vectors that represents the edges of the graphs.

For each element and each possible edge we save the coordinates of the vector for each frame of the scene. One position vector for each element and M free vectors for the edges, where $M = \frac{N*(N-1)}{2}$.

To describe the spatial relation between elements j and j in the frame t, there is a vector called $v_{i,j,t}$ where $r_{i,j,t} = |r_{i,j,t}|^{\theta_{i,j,t}}$, with $|r_{i,j,t}| = \sqrt{(x_{j,t} - x_{i,t})^2 + (y_{j,t} - y_{i,t})^2}$ and $\theta_{i,j,t} = \tan^{-1}\frac{(y_{j,t}-y_{i,t})}{(x_{j,t}-x_{i,t})}$.

Fig. 2 Graphic representation of a system with six elements and three groups

4 A Structured Representation to the Group Behavior Recognition Issue

In most the case of group behavior recognition it is difficult to identify any individual action and role played by each element of the group. Otherwise, tracking a group of elements in computer video analysis could provide the positioning of the elements in the scene.

There are a lot of systems providing the location of elements of a group, like GPS (outdoor), UWB (indoor), tracking, etc.

Behavior recognition based on the positioning of each element of the group could provide results without knowing the activity carried out by the elements of the group.

We propose a structured representation composed by three matrix called R,A and S.The first one save the geometrical data of the elements in time (the position vectors and free vectors),while the second one represents the information about the semantic of the scene, composed by the number of groups founded and their makeup, and the third one represents the shape of each graph of the scene.

This structured representation contains the information about the position of each element of the scene, the information about the spatial relations between the elements, and the groups shape information. It is importance to notice that the last one information could be representative of the behavior of the group in many different domains, like group sports, surveillance, defense, ethology, etc.

4.1 Geometrical Information

All the geometrical information about the elements in the scene, and its relations is saved on matrix R, this information is used to construct the shape matrix.

Matrix R is a three dimensional matrix with the information of the location of each agent and the free vector that represent each possible edge presented at the scene.

A scene with N elements has $M = \frac{N*(N-1)}{2}$ possible edges that must be saved, and N position vectors.

Each vector of the matrix has two or three components (it is depending on at the casesa 3-D scene or 2-D scene).

The first ones N vectors represent the location of the N elements, and the next M vectors represent the edge of the graphs.

The R matrix has one row for each frame of the scene and N + M columns.

$$\begin{pmatrix} |r_1|^{\theta_1} & \cdots & |r_N|^{\theta_N}|r_{1,2}|^{\theta_{1,2}} & \cdots & |r_{N-1,N}|^{\theta_{N-1,N}} \\ \vdots & \ddots & & \vdots & \\ |r_1|^{\theta_1} & \cdots & |r_N|^{\theta_N}|r_{1,2}|^{\theta_{1,2}} & \cdots & |r_{N-1,N}|^{\theta_{N-1,N}} \end{pmatrix}$$

4.2 Semantics Information

The semantics information represents the associations between the elements of the scene to perform groups. One element could be part of many groups, and could be many groups.

This information makes it possible to create different associations between elements to grasp better the semantics context.

This semantics information is saved in a binary matrix with one row for each group, and one column for each element.

The matrix can only contain zeros or ones, which represent if this element forms part of the graph.

For example, in a scene with nine elements, and two groups, the matrix A could be like this: $A = \begin{pmatrix} 1 & 1 & 1 & 1 & 1 & 0 & 0 & 0 & 0 \\ 0 & 0 & 0 & 0 & 1 & 1 & 1 & 1 & 1 \end{pmatrix}$. This matrix shows that there are two graphs, the first one composed by the elements: 1, 2, 3, 4, and 5; and the second one composed by the elements: 5, 6, 7, 8 and 9.

4.3 Shape Information

Matrices S define the shape of the graphs, there are one matrix for each graph (row of the A matrix).

Each S matrix has a number of T rows, and one column for each possible edge of the graph. The number of possible edge depends on the number of elements that composed the graph ($M_g = \frac{N_g*(N_g-1)}{2}$).

Elements of S have two components: relative distance (d) and direction (γ).

Relative distance is calculated by the formula $d = \left\lceil \frac{|r_{ij}|*8}{|r_{max}|} \right\rceil$. Where r_{ij} is the distance between the elements i and j, and r_{max} is the maximum distance between any elements of the graph. By definition d is a natural number between 1 and 8.

Direction between two elements of the graph is defined by the formula $\gamma = \left\lceil \frac{|\theta_{ij}|*8}{\pi} \right\rceil$. Where θ_{ij} is the angle between the elements i and j, and. By definition γ is a natural number between 1 and 8. It is important to perceive that in spite of the graphs are not directed, to construct the reduced graph we have to distinguish between the same directions with different sense. So the possible directions arecovered between -π and π radians.

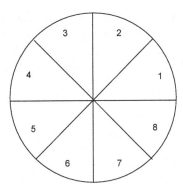

Fig. 3 Directions code

Each S matrix has the edges of the graph that defines its shape. If some edge of the element ihave the same direction that another one and it is longer than the previous one, then this edge is not added to the matrix. One negative value (-1) is added in this position.

Figure 4 shows the construction process: the shortest edge of the first element is added in (a). Then the second shortest is also added in (b). In (c) there is a shorter edge with the same direction (2) and the edge is not added. The process is repeated until all the elements are checked (d), (e) and (f).

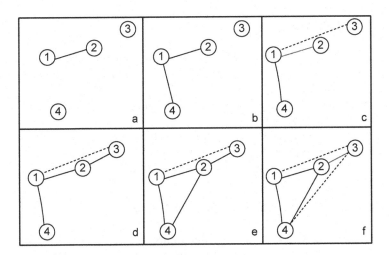

Fig. 4 Construction process

The matrix below shows the S matrix of the graph in the Fig. 4. First row represents the graph at the instant t = 0, and last row shows the graph at the instant t = T.

Fig. 4 shows an example of a graph with five elements. In the first frame the edges between the nodes 1-3, 3-4 and 3-5 are not defined because they have the same directions (and they are longer) than the edges 1-5, 3-2 and 5-2.

Then, in the frame T the graph's shape has changed, there are new relevant edges (like 3-5 and 3-4) and some relative distance have also changed.

$$\begin{pmatrix} 4^6 & 6^8 & 3^4 & -1^5 & 3^2 & 4^1 & 5^3 & -1^1 & -1^3 & 4^5 \\ \vdots & \vdots & \vdots & \vdots & \vdots & \vdots & \vdots & \vdots & \vdots & \vdots \\ 6^6 & 5^8 & 4^4 & -1^5 & 4^5 & 6^8 & 5^2 & 6^1 & 7^3 & 5^5 \end{pmatrix}$$

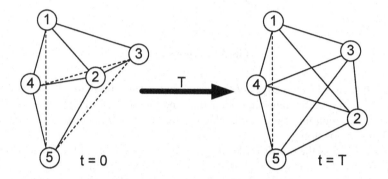

Fig. 5 Example graph

5 CVBASE06 Dataset

In the previous section, we define a structured representation for group behavior recognition. These paragraphs are going to show a practice application for the European handball domain.

We have used a dataset called CVBASE06 dataset, from: http://vision.fe.uni-lj.si/cvbase06/downloads.html.

CVBASE06 dataset (9)has three sports: squash, basketball and European handball. We have selected the handball dataset because it has the positioning of the player and the action played by the team at each moment.

5.1 European Handball CVBASE06 Dataset

The handball CVBASE06 dataset part has information about the players (their position) and about the teams (the action played), distributed in six files.

There are ten minutes of video recorded with three cameras (A, B and C), all of them synchronized. Cameras A and B record the match from the top of the sports hall, and camera C record from the typically TV view.

The video has 25 fps, so there are 15000 frames. Each frame has the position of each player of a team, and the group action played by this team. All of this information is structured on plain text files.

Each group action start when finish the previous one, so in each frame there is one, and only one, group action, and all the frames have one defined group action.

A European handball team is composed by seven players, one of them is a special player called goalkeeper, whose only can run around from his own goal area.

5.2 Active Attack and Passive Attack

There are two different types of attack in European handball, one of them aims to score quickly, and the other one tries to keep the ball. The first one is called an active attack, and the second one is called passive attack. Normally one attack starts with a passive attack, and end with an active attack in the same move.

The two different types of attack are labeling with: "nfan" and "nfpn" respectively.

For this domain we have define the A matrix to create a group with four players of the attacking team.

The figure 6 shows the S matrix for two sequences of the video, in the first one (a) we can see a passive attack, and in the second one we can see a active attack (b).

The "frame" axis represents the evolution of time, in frames. The "edge" axis represents each edge of the graph, numbered on (c). The (d) shows a frame of the scene, with the elements of the graph colorized.

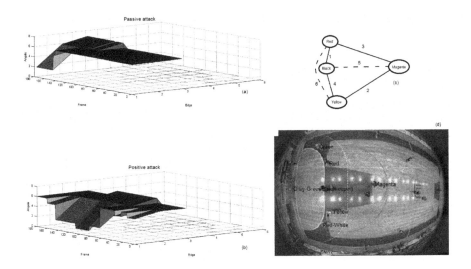

Fig. 6 Passive/Active attack

Different dynamism of the two types of attack is reflected on the "frame" axis. In the first attack, there are few changes, but in the second one there are a lot of changes.

In the fig. 6 (a) and (b) also we can see that the graph constructed with the S matrix. In this graph the algorithm has deleted the edges 5 and 6, edges 1, 2, 3 and 4 have all the information about the formation of the group.

The different dynamism of the two types of attacks is also showed in the probability of change of the S matrix throughout the time.

Fig. 7 shows the probability of two different values consecutive in S matrix. Probability of change angle or distance in active attack (nfan) is generally greater than probability of change in passive attack.

So there are reasons to think that the matrix representation keeps the essential information and it is more compressed model.

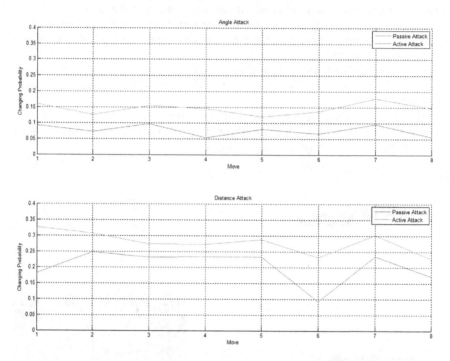

Fig. 7 Changing Probability

6 Conclusions

There are a lot of papers related with the behavior recognition, but most of them are related with one single element. This paper had focused on the group behavior recognition, where a group of elements compose a group and have a group behavior.

In the group behavior recognition the researches have followed two ways approach: logical and geometrical. Because there are a lot of systems providing the location and tracking of elements of a group, this paper had focused on the geometrical approach.

This approach had reduced the number of relations without loss information about the formation to realize the rezoning process. This approach is based in a novel structured representation of the important relations between the elements of the graphs.

This structured representation save the information about the formation of the group, wasting less space.

One of the assumption makes is that the formation information of the group (and its dynamics) is representative of the group behavior. This assumption could be applied to many different domains like: defense, etholy, group sports, etc.

Acknowledgements. This work was supported in part by Projects CICYT TIN2008-06742-C02-02/TSI, CICYT TEC2008-06732-C02-02/TEC, CAM CONTEXTS (S2009/ TIC-1485) and DPS2008-07029-C02-02.

References

1. Aggarwal, J., Cai, Q.: Human motion analysis: a review. Computer Vision and Image Understanding, 428–440 (1997)
2. Bobick, A., Wilson, A.: A state-based approach to the representation and recognition of gesture. In: PAMI, pp. 1325–1337 (2002)
3. Moeslund, T.B., Kruger, V., Hilton, A.: A survey of advances in vision-based human motion capture and analysis. Computer Vision and Image Understanding, 90–126 (2006)
4. Ryoo, M.S., Aggarwal, J.K.: Recognition of High-level Group Activities Based on Activities of Individual Members. WMVC (2008)
5. Khan, S.M., Shah, M.: Detecting Group Activities using Rigidity of Formation. ACM, Singapore (2005)
6. Ruonan, L., Rama, C., Shaohua, K.Z.: Learning Multi-modal Densities on Discriminative Temporal Interaction Manifold for Group Activity Recognition, New York (2009)
7. Perse, M., et al.: A Trajectory-Based Analysis of Coordinated Team Activity in a Basketball Game, Ljubljana (2009)
8. Ramos, F., Ayanegui, H.: Tracking behaviours of cooperative robots within multiagent domains, Tlaxcala (2010)
9. Pers, J.: CVBASE 2006 Dataset: A Dataset for Development and Testing of Computer Vision Based Methods in Sport Environments, Ljubljana (2005)
10. Intille, S., Bobick, A.: Representation and visual recognition of complex multi-agent actions using Belief networks. In: Burkhardt, H., Neumann, B. (eds.) ECCV 1998. LNCS, vol. 1407, Springer, Heidelberg (1998)

Opportunistic Multi-sensor Fusion for Robust Navigation in Smart Environments

Enrique Martí, Jesús García, and José M. Molina

Abstract. This paper presents the design of a navigation system for multiple autonomous robotic platforms. It performs multisensor fusion using a Monte Carlo Bayesian filter, and has been designed to maximize information acquisition. Apart from sensors equipped in the mobile platform, the system can dynamically integrate observations from friendly external sensing entities, increasing robustness and making it suitable for both indoor and outdoor operation. A multi-agent layer manages the information acquisition process, making it transparent for the core filtering solution. As a proof of concept, some preliminary results are presented over a real platform using the part of the system specialized in outdoor navigation.

Keywords: multi agent, sensor fusion, positioning, indoor/outdoor navigation, particle filter.

1 Introduction

In the last years we have seen several implementations of autonomous robotic platforms doing simple works or assisting humans in theirs. These realizations have taken place in environments as disparate as hospitals [1] or factories[2], and, in spite of their relative simplicity, they can be seen as an advance of a future in which robotic workers will be massively used in more complex tasks. In view of such potential scenario, both self-location and navigation are problems of the uttermost importance for achieving continued, fail-proof operation.

This work aims to introduce a simple but robust architecture for combined indoor/outdoor navigation through sensor fusion technology, where the information provided by on-board sensors is aligned with heterogeneous external references [3] coming from different sources.

The core of the navigation solution is implemented as a Sampling Importance Resampling (SIR) Particle Filter (PF) [4] with loose coupling integration of

Enrique Martí · Jesús García · José M. Molina
Group of Applied Artificial Intelligence, Universidad Carlos III de Madrid,
Av. de la Universidad Carlos III, 22, 28270 Colmenarejo, Madrid, Spain
e-mail: emarti@inf.uc3m.es, jgherrer@inf.uc3m.es,
molina@ia.uc3m.es

J.M. Molina et al. (Eds.): User-Centric Technologies and Applications, AISC 94, pp. 59–68.

received information. Apart from being capable of dealing with the non-linear relation between internal sensors and absolute external references, this Bayesian inference tool requires minimal efforts for integrating new types of sensor measures –to the point of not needing previous knowledge about the sensor: just the type of data it is providing and a description of the associated uncertainty.

Our proposal achieves accuracy and reliability through redundancy, considering the set of sensors to be fused as a changing entity. Apart from the typical internal devices –IMU, laser, GPS, odometry–, the robotic platform is able to obtain additional information from external entities, such as Ultra Wide Band sensor networks or external video-based trackers.

Works on navigation through multisensor fusion usually define architectures and algorithms specifically tailored for a selected set of sensors, as in the cases of tight coupling or feature based navigation[5].Those specialized approaches can take advantage of the existing synergies between different sensing technologies, so that the final result is more stable, accurate or computationally affordable. Nonetheless, it is more sensitive against changes in the set of sensors: integrating new technologies can be a difficult task, and if an existing device suffers a temporal outage, the system could even be unable to continue its normal function.

Regarding the decision of integrating external sensors, some researchers point to the convenience of pure standalone robots arguing that ad-hoc sensor networks are expensive, could be unavailable, and same or better results can be obtained using internal devices [6]. Our proposal is, however, based in the fact that communication and sensor networks are more common every day. Moreover, multiple robotic platforms coexisting in an environment can be simultaneously benefited from them. This means that the proportional cost of such installation has to be divided and the amount of internal sensors on each mobile unit can be cut back – with the subsequent expenses reduction.

The central part of this document contains a detailed description of the proposed system, while the last sections report the results obtained with a first partial implementation capable of outdoor navigation using an Inertial Measure Unit (IMU) and a GPS device. A section with some conclusions and projected future work closes the document.

2 Description of the Proposed System

The system can be split into two parts: the sensor fusion process inside a single robotic mobile platform, and the cooperative network formed by several of these devices together with an intelligent environment. Let us begin with the internals of autonomous mobile platforms.

2.1 Fusion Architecture for a Single Platform

From the architectural point of view, the navigation system of each robot is organized in layers. Each tier plays a different role in the process of acquiring

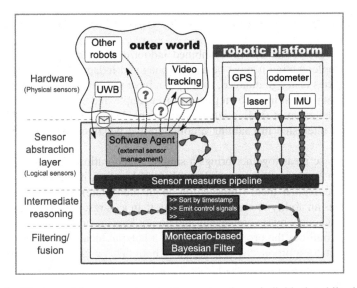

Fig. 1 Architecture of the proposed navigation system for an individual mobile platform

information, and transforming it into something useful for the sink of final data: the filtering algorithm. Figure 1. contains a schematic view of the system.

Our solution follows the design principle stated in the introduction of being robust against sensor outages. The selected architecture includes mechanisms for this purpose from the very beginning of the fusion process –data acquisition–to the filtering solution.

Right at the top of the architecture is the *Sensor abstraction layer*. This level manages communication with both internal (on-board) and external real sensing devices, providing lower levels with a view of logical data sources with unified access interface. Thus, this abstraction level supports online sensor addition and removal, while isolating inferior processes from nasty details and the consequences associated to such changes.

The *Intermediate reasoning layer* receives bare measures from sensor abstraction level and adapts it to the filter various necessities. Our first implementation is restricted to basic operations required by lower layers, assorting sensor readings according to the time when they were generated. However, it is intended to support advanced processes of high-level reasoning, as the application of fuzzy reasoning to detect sensor malfunction.

In the bottom layer, the *Filtering solution* has been chosen to have a reactive working profile: incoming sensor data is integrated as it is fed by the superior layer. It does not impose the presence of specific sensors nor a predefined schedule/order for data arrival.

Following subsections cover the multi-agent system proposed for exchanging sensory information, and the selected filtering solution. Sensor abstraction and intermediate reasoning layers do not have dedicated sections because of their simplicity at current development stage.

2.2 Multi-agent System for Sensory Information Exchange

Being the target of our system combined indoor/outdoor navigation, the mobile platforms implementing it must be autonomous and independent: the presence of external entities is not guaranteed, so they can have to operate in standalone mode.

Central data fusion in the own platform is a simple and effective solution. Nonetheless, in the event of finding other entities capable of providing useful information, we are interested in enabling a collaborative behavior to extend navigation capabilities.

Putting these ideas together brings a simple collaboration scenario, where individuals (each Agent) have compatible goals (self-location, navigation) but lack the ability to accomplish them (information about their state). Information is interpreted here as an ability to reach a goal rather than as a resource, because it can be used simultaneously by several agents and do not lose its value when used.

Figure 2. outlines the proposed system as a heterarchical scheme where agents can be information producers, consumers or both. This architecture has been preferred over any type of hierarchy because it gives the system the desired flexibility and robustness.

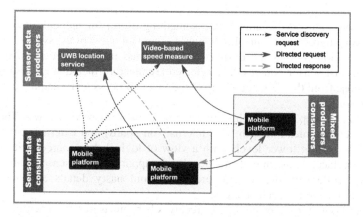

Fig. 2 Schematic example of Multi Agent system for enabling external sensor fusion

Consumers are in charge of asking for new services and requesting sensor measures, to which producer agents can respond or not. In spite of the relative inefficiency of the approach, it offers many advantages in this dynamic scenario: in first place it does not require the existence (and persistent accessibility) of infrastructures as service directories or mediation agents. The second main advantage is that each agent concentrates all the information about its own location/state, so it is the best suited entity for deciding which type information it needs and when to ask it.

Agents will use a simple ontology based in GONZ[7] for information exchange. This ontology has been modified and extended so that it can be used with arbitrary

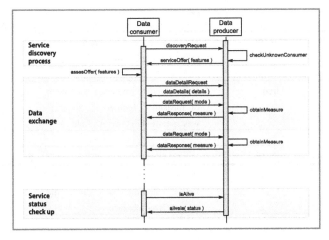

Fig. 3 Interaction diagram showing the caracteristical life cycle of a relation betweena consumer agent (mobile platform) and a producer (external sensor)

types of sensors –not restricted to location services–and includes the additional information required by the fusion algorithm employed in the mobile platforms.

Four pairs of messages are defined for the three types of interaction between agents: service discovery, data exchange and service status check. Almost all communication processes follow a client-server model.

Each action involves the transmission of two messages, the first one making a request (information consumers) and the other responding the petition (information producers). Figure 3. contains all the type of messages available, that we proceed to enumerate and describe. By default (not in the tables) all messages include the unique ID of the sender, the unique ID of the recipient if it is a directed message, and a timestamp.

discoveryRequest	Field	Description	Sample values
(type: broadcast)			

serviceOffer	Field	Description	Sample values
(type: directed)	measureType	Measureable physical magnitudes	PositionXYZ, SpeedXY

dataDetailRequest	Field	Description	Sample values
(type: directed)			

dataDetails	Field	Description	Sample values
(type: directed/broadcast)	error	Description of error distribution	
	samplingFreq	Maximum frequency with which the external sensor can return measures, in Hz	0.01, 5

dataRequest	Field	Description	Sample values
(type: directed)	mode	How the consumer wants to receive measures: only under request, or periodically	{request, stream}
	freq	Desiredupdate frequency for stream of measures. Ignored if "request" was specified in previous field	3

dataResponse	Field	Description	Sample values
(type: directed)	value	Array of values according to the field measureType in the serviceOffer message	[0.5, 3, -1.2]

isAlive	Field	Description	Sample values
(type: directed)			

aliveIs	Field	Description	Sample values
(type: directed)	status	Availability of the external sensor to attend requests at this moment	{ready, busy}

The communication scheme has a very simple design, avoiding intricate interaction patterns. Nonetheless, there are two important remarks to be made:

- dataDetails message is the only one that can be sent by information producers without direct request. Changes in the maximum update frequency or in the measure uncertainty can trigger its broadcasting
- The error distribution description in dataDetails message follows its own syntax. Its contents can vary from analytical (multivariate Gaussian distribution: mean and covariance matrix) to sample-based descriptions as that used in [8].

2.3 Monte Carlo Based Bayesian Filter

For tracking a mobile platform and integrating all the available information about it, we need a filtering algorithm capable of dealing with non linear models – prediction and errors–. The algorithm must also be flexible enough to overcome the difficulty of unstructured data presentation.

The most common filtering algorithms use Bayesian inference to estimate the state of a partially observed system (in our case, position and dynamics of a mobile target). The uncertainty about true state makes necessary to store the belief as a probability distribution, so that at each time the filter can estimate which is the most probable state according to the available information. This probability distribution changes with time. It can be adapted using a prediction model that describes system dynamics, and incorporating some occasional measurements providing information about the real state.

Some techniques, as the Kalman Filter (KF) or the Extended Kalman Filter (EKF) [9] assume that all uncertainties have Gaussian distribution, and store the state probability distribution as another Gaussian. Thanks to that simplification, they obtain a matrix-based analytical formulation of the filtering process that can be calculated fast (and is optimal if the assumptions are true).

Nonetheless, if system dynamics obey a highly non-linear model or uncertainties are far from being Gaussian, then these techniques deliver poor performance. A PF[4][10] is a Monte Carlo algorithm capable of dealing with such non-linear non-gaussian scenarios.

Some techniques as the Rao-Blackwellized [11] or the Unscented Particle Filter [12] are known to be more effective than plain PF, but their formulations have subtle details that make more difficult the dynamic integration of heterogeneous sensors. Instead, enabling a Particle Filter for working with a new sensor is as simple as specifying: (a) the sensor error model, (b) function relating state vector with sensor measures (only for sensors providing direct evidences of state), and (c) integration with the prediction model (only if the sensor provides control inputs).

The critical part of our system lays in the procedure for integrating external sensors, about which the mobile platform does not have prior information. Nonetheless, PF can obtain the required data elements as follows: (a) is provided by the own sensor through dataDetails message; (b) is specified by measureType field in serviceOffer message. (c) is not contemplated in our scheme, meaning that external sensors are restricted to provide absolute references.

3 Outdoor Navigation Experiments

This last section contains some results implemented over a real platform, intended to assess the suitability of the proposed system under different conditions.

The experiments of this work are based in a GUARDIAN rover from Robotnik corporation [13]. It features a wide range of sensors, including but not limited to odometry, laser ranging, inertial navigation and a video camera. However, the experiments are focused in outdoor navigation and are based in the data obtained by an Inertial Measure Unit (IMU) and a GPS device.

The IMU is a InertiaLink 3DM-GX2 unit [14] containing triaxial accelerometer, gyro and magnetometer. Only accelerometer and gyroscope readings will be used. The global position sensor is a Novatel® OEMV-1G differential GPS [15]. It is compatible with Satellite Based Augmentation Systems as EGNOS, though it has been operating on single point L1 mode for these experiments –accuracy of 1.5 m (RMS) in ground positioning.

The robotic platform is equipped with an embedded computer for high-demanding computing tasks. It also allows the integration of sensing and control hardware through the Player/Stage architecture [16].

External references are obtained only by means of a GPS –internal sensor–, so the platform will be subject to impoverished sampling frequencies and outage periods to mimic adverse conditions that can arise using external sensors.

3.1 Fusion Performance

This section presents the results of executing the proposed experiments with simulated and real trajectories. Special attention will be put in unexpected behaviors and other unusual effects, to help refining the system in future developments.

A stadium-shaped trajectory was used to calculate a general view of the overall filter performance. Receiving GPS measures at 1 Hz, the PF can easily obtain an average 40% improvement over bare observations, as shown in Figure 4.

The last test use a circular trajectory to see how the system performs when only low-accuracy positioning is available in reduced spaces and the availability of external references is restricted. Figure 5. shows the obtained filtered trajectory. In

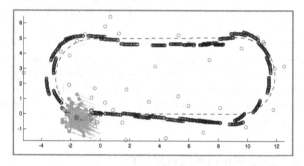

Fig. 4 In a stadium-shaped trajectory the PF improves GPS accuracy an average 40% (from meter-precision to 0.6 m).

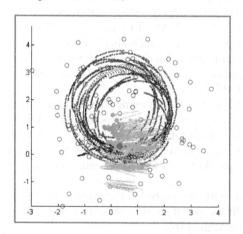

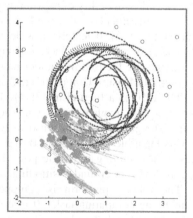

Fig. 5 Filter results using low accuracy positioning in reduced spaces

Fig. 6 Same experiment using a 0.1Hz GPS signal

spite of its positioning error below 0.55 m –even better than the stadium case–,the motion pattern is erratic, with large orientation errors. Figure 6. Shows that the prediction is much worse when GPS measures arrive each 10 seconds. Particle population is largely scattered through state space, but the filter do not diverge.

4 Conclusions and Future Work

A system for robust navigation in both indoor and outdoor environments has been presented. Apart from being able to operate in standalone mode, it can interact with smart environments to take advantage of an extended sensory capability.

The architecture of the interaction system has been defined to support the simultaneous load of many robotic platforms, while being capable of dealing with any kind of failure. Moreover, mobile platforms using the proposed navigation system can integrate external sensors straightforwardly, without requiring any kind of configuration.

Smart environments including this system into their functionality only have to implement the multiagent system as an independent module. Its design is focused in not being intrusive and making a rationale use of resources, so that early functionality is not affected by the execution of the sensory services.

The platform has been subject to some preliminary tests, resulting in acceptable results even under conditions of reduced sensor availability. Future work include the integration of magnetometer, odometer and laser sensor as internal devices, and two external positioning systems: Ultra Wide Band and video tracking.

Some of these sensors have been included in other works [17] to achieve exceptional accuracy levels that we should be able to reproduce with reduced effort.

Acknowledgment

This work was supported in part by Projects CICYT TIN2008-06742-C02-02/TSI, CICYT TEC2008-06732-C02-02/TEC, CAM CONTEXTS (S2009/ TIC-1485) and DPS2008-07029-C02-02.

References

[1] Takahashi, Y., Kohda, M., Kanbayashi, Y., Yamahira, K., Hoshi, T.: Tray Carrying Robot for Hospital Use. IEEE, Los Alamitos (2005)
[2] Wulf, O., Lecking, D., Wagner, B.: Robust Self-Localization in Industrial Environments based on 3D Ceiling Structures. IEEE, Los Alamitos (2006)
[3] Cannon, M.E., Nayak, R., Lachapelle, G., Salychev, O.S., Voronov, V.V.: Low-Cost INS/GPS Integration: Concepts and Testing. The Journal of Navigation 54, 119–134 (2001)
[4] Gordon, N.J., Salmond, D.J., Smith, A.F.M.: Novel approach to nonlinear/non-Gaussian Bayesian state estimation. IEE Proceedings F Radar and Signal Processing 140, 107–113 (1993)

[5] Huang, L., Barth, M.: Tightly-coupled LIDAR and computer vision integration for vehicle detection. IEEE, Los Alamitos (2009)

[6] Zampieron, J.: Self-Localization in Ubiquitous Computing using Sensor Fusion (2006)

[7] Gil, G., Berlanga de Jesús, A., Lopéz, J.M.: Multi-sensor and Multi Agents Architecture for Indoor Location. Distributed Computing and Artificial Intelligence - Advances in Soft Computing, 309–316 (2010)

[8] Rosencrantz, M., Gordon, G., Thrun, S.: Decentralized Sensor Fusion with Distributed Particle Filters. In: Proc. of UAI (2003)

[9] Welch, G., Bishop, G.: An Introduction to the Kalman Filter (1995)

[10] Arulampalam, M.S., Maskell, S., Gordon, N.: A tutorial on particle filters for online nonlinear/non-Gaussian Bayesian tracking, vol. 50, pp. 174–188 (2002)

[11] Gustafsson, F., Gunnarsson, F., Bergman, N., Forssell, U., Jansson, J., Karlsson, R., Nordlund, P.-J.: Particle filters for positioning, navigation, and tracking. IEEE Transactions on Signal Processing 50, 425–437 (2002)

[12] Merwe, R.V.D., de Freitas, N., Doucet, A., Wan, E.: The Unscented Particle Filter.

[13] Automation, R.: Guardian Mobile Robot datasheet (2010)

[14] MicroStrain, Inertia-Link ® 3DM-GX2 Datasheet (2008)

[15] Novatel, OEMV-1G GNSS Card Datasheet (2010)

[16] Gerkey, B.P., Vaughan, R.T., Howard, A.: The Player/Stage Project: Tools for Multi-Robot and Distributed Sensor Systems, pp. 317–323 (2003)

[17] Cheng, Y., Crassidis, J.L.: Particle Filtering for Attitude Estimation Using a Minimal Local-Error Representation. Journal of Guidance, Control, and Dynamics 33, 1305–1310

Automatically Updating a Dynamic Region Connection Calculus for Topological Reasoning

Miguel A. Serrano, Miguel A. Patricio, Jesús García, and José M. Molina

Abstract. During the last years ontology-based applications have been thought without taking in account their limitations in terms of upgradeability. In parallel, new capabilities such as topological sorting of instances with spatial characteristics have been developed. Both facts may lead to a collapse in the operational capacity of this kind of applications. This paper presents an ontology-centric architecture to solve the topological relationships between spatial objects automatically. The capability for automatic assertion is given by an object model based on geometries. The object model seeks to prioritize the optimization using a dynamic data structure of spatial data. The ultimate goal of this architecture is the automatic storage of the spatial relationships without a noticeable loss of efficiency.

Keywords: RCC Automatic Assertion, Dynamic Topological Relationships, Ontology-based Application, Ontology-centric Application.

1 Introduction

Knowledge approaches have always used ontologies to conceptualize and organize interpretations of the real world. Since its conception ontologies were designed for reasoning with preset information. This information also had the condition that must not undergo intensive updates. In the last years, researchers have been developing ideas which assumes that ontology applications are prepared to automatically populate a knowledge base or to bring about changes in a large amount of data at the same time. Nowadays none of these functions can be done maintaining an acceptable performance from a given number of updates.

Lately, interest and necessity to carry out the semantics of the representation between domain objects with spatial characteristics is taking a greater presence in knowledge applications. Spatial representations are limited by the description

Miguel A. Serrano · Miguel A. Patricio · Jesús García · José M. Molina
GIAA Computer Science Departament, Universidad Carlos III de Madrid, Avda. de la
Universidad Carlos III, 22, 28270 Colmenarejo, Madrid, Spain
e-mail: {miguel.serrano, jesus.garcia,
miguelangel.patricio, josemanuel.molina}@uc3m.es

J.M. Molina et al. (Eds.): User-Centric Technologies and Applications, AISC 94, pp. 69–77.

logic standard used in ontologies because the definition of the spatial logic is very wide and complex and also is a not very efficient alternative in terms of calculation time. To overcome this limitation spatial representations are achieved through the Region Connection Calculus (RCC) logic formalism. RCC is a theory [1] for qualitative spatial representation and reasoning. This theory provides a formalism which allows inferring implicit knowledge which is hidden but is deductible from explicit knowledge declared in concrete statements and a set of qualitative spatial relationships to describe the relative location of spatial objects to one another.

The loss of performance when updates occur intensively and the increased use of RCC could result in a collapse in ontology applications. This paper presents an architectural model for solutions which combines dynamic spatial relations with the need of automated RCC assertions and updates. As was previously introduce this kind of approach needs a compromise between the automatic storage of dynamic RCC data with the minimal loss of performance in terms of time.

To obtain automatic functionality the spatial individuals represented as geometric entities have to be instantiated in a Euclidean planar linear geometry model. This model must discover, maintain and provide the relationships between these geometries. All these tasks are carried out through an analysis performed after each update of the geometries in the model. Spatial relationships are only updated when the geospatial situation of the objects causes changes in the general object organization.

To keep the spatial relationships updated is necessary to check all the instantiated objects pairwise. To accomplish this, it would have to perform a comparison of quadratic complexity at every moment. This task may cast down a good overall performance. For example in a video tracking system could be necessary check, at every frame, the number of tracked and static geometries relationships. To resolve this kind of situations we will show an approach which uses an auxiliary data structure capable to reduce the difficulty of the search to logarithmic complexity.

Spatial relationships from the geometric model are finally stored to a RCC module as qualitative spatial relationships. Knowledge is shared between the knowledge base and the RCC. RCC module contains the topology relationships of the spatial individuals who belong to the knowledge base. Access to knowledge is guaranteed since knowledge base individuals can be easily identified in the RCC module.

The paper is organized as follows. In Section 2 approaches related with the RCC are studied briefly; Section 3 the dynamic RCC automatically update overall architecture is presented; Section 4 shows an application example to solve a typical tracking problem applying our solution; Section 5 explains the conclusions obtained and the future work.

2 Spatial, Temporal and Topological Approaches

The problem of the spatial relationships between objects is very common in the spatial databases field. Unfortunately, there are no designed architectures for the automatic and optimized assertion of spatial information in the knowledge bases field.

Spatial databases follow a similar approach; receive and store spatial data in a tree trying to optimize the objects searches. Spatial databases have the disadvantage that they cannot do reasoning with objects; however knowledge approaches are thought to do it.

In recent years the joint use of the spatial and temporal dimensions combined with ontology reasoning is growing. Nowadays there are ontology-based proposals which include spatial reasoning with dynamic geometries [2]. There are also trends which combines knowledge system architecture with a representation of the RCC family tree in the Ontology Web Language (OWL) [3]. Even there are architectures with reasoning systems for visual information fusion [7].

The most remarkable standards in the topology field are developed by the Open Geospatial Consortium previously known as the Open GIS Consortium. These are the result of an agreement process to develop publicity available interface standards. OpenGIS standards support interoperable solution ranging from web wireless to the location-based services and mainstream IT [12].

RCC theory can be seen as a topology standard in the knowledge bases field. RCC spatial representation uses topological relations which are compliant with the OpenGIS standard [4]. There are publications dealing with the mapping of existing relationships in the OpenGIS standard to the RCC theory [1]. The RCC is an axiomatization of certain spatial concepts and relations in first order logic. The basic theory assumes just one primitive dyadic relation: $C(x, y)$ read as "x connects with y". Individuals (x, y) can be interpreted as denoting spatial regions. The relation $C(x, y)$ is reflexive and symmetric.

Of the defined relations, Disconnected (DC), Externally Connected (EC), Partially Overlaps (PO), Equal (EQ), Tangential Proper Part (TPP), Non Tangential Proper Part (NTPP), Tangential Proper Part Inverse (TPPi) and Non Tangential Proper Part Inverse (NTPPi) have been proven to form a jointly exhaustive and pairwise disjoint set, which is known as RCC-8. Similar sets of one, two, three and five relations are known as RCC-1, RCC-2, RCC-3 and RCC-5.

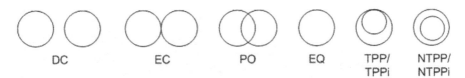

Fig. 1 RCC-8 relations.

Currently there are implementations of this theory for two of the most powerful systems of reasoning based on ontologies, Pellet [5] and RACER.

3 Overall Architecture

Dynamic RCC system presented in this paper is based on three modules; a container with individuals with geometric features which provides the entry data to the main module, a synergistic module composed of an object model and an auxiliary data structure, and a Region Connection Calculus where the qualitative spatial relationships between the objects are stored. The overall architecture of the proposed framework is illustrated in Fig. 1.

Since this architecture is thought for ontology-based systems, geometric individuals are initially stored in a knowledge base. This knowledge base contains static objects whose position does not change and dynamic objects whose position is altered over time. All the geometric representations of these objects are used as input to the main module and more specifically to the sub-module that contains the object model. The geometric representations are instantiated in the object model in two cases; if they do not exist previously in the model or if they were already instantiated but its position or size has been modified regarding the last update. As a result of these changes it is necessary to perform a full topological analysis of the relations between the new instantiated geometries with the existing geometries. This type of analysis has a quadratic complexity, since it is necessary to compare each geometry with the others.

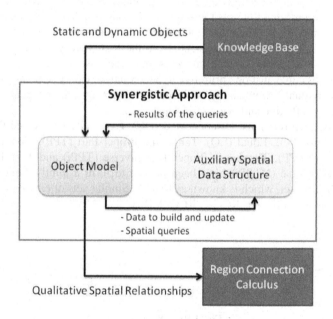

Fig. 2 Overall Architecture.

This difficulty can be translated into a decrease of the system efficiency. To avoid this problem, the number of checks between geometries has to be reduced. To achieve this is necessary to reduce the number of geometries selecting only those are candidates to modify the spatial relations of each geometry. By applying an auxiliary data structure, these candidates, who make up a subgroup inside of the geometries total group, can be determined.

Once the auxiliary data structure returns the candidates, the object model can be fully updated. The qualitative spatial relationships which have changed from the previous state are then updated on the RCC system. The RCC characteristics of the objects are independent of the knowledge base therefore it is not necessary to make changes in the instances of the ontologies.

3.1 Object Model

Synergistic approach is composed of two sub-modules that work in a complementary way; an object model and an auxiliary data structure. The object model is a system that represents spatial objects in a Euclidean plane and is capable of obtaining quickly 2D spatial relationships between objects. Besides implementing the geometric model its operations could define the OpenGIS Simple Features standard [13]. This standard specifies digital storage of geographical data with spatial and non spatial attributes and defines a set of spatial operators.

Sometimes, building spatial data structures is necessary to know the objects speed and the smallest enclosing rectangle for their geometric representations. Object speed data can come from the knowledge base or from the object model. Although the speed may not be considered a spatial data, the OpenGIS standard implemented by the object model ensures the possibility of include non-spatial attributes in its model. To find the smallest rectangle enclosing just need to find the minimum area which limit points of the geometry.

OpenGIS spatial representations and RCC are not compatible. The output from the object model must be mapped from the standard features of the OpenGIS to the RCC-8 base relations. This kind of translation has been previously proposed in [8].

As mentioned earlier the object model automatically instantiate all the new or updated geometries corresponding to the spatial representations of objects from the knowledge base. To discover the spatial relationships between geometries should be carried out a topological analysis of these new instances. This analysis leads to a quadratic order check because it is necessary to check the new items with everyone else in every moment of time. The exactly number of total checks is $N*(N-1)/2$.

In a scenario in which mobile objects move in a consistent manner, spatial relationships change between objects that are close in consecutive time instants. For this reason, it is necessary to do the topological analysis to geometries that are physically close to the geometry whose relationships are being analyzed. A structured topological hierarchy that can change over time may therefore improve the performance of this approach. Starting from basic information on the situation of the geometries it is possible to build a spatial data structure that maintains a hierarchical topological sort on the Euclidean space and supports spatial queries. Performing spatial queries on this structure is possible to reduce the amount of geometries involved in the object model analysis. To do so each query has to return for each geometry which candidate geometries can change their spatial relationships. The results of the queries to the spatial data structure are the candidate geometries who were next to the geometry whose relations were evaluated in time when the query was performed.

3.2 Auxiliary Spatial Data Structure

This spatial data structure is not predefined by the architecture, however it is recommended that follow several conditions; must be capable of defining a spatial hierarchy throughout the time, should be able to handle the overlap between objects and its operations must have a complexity lower than the quadratic.

It is strongly recommend that search, deletion, insertion and update operations do not carry a high overhead because the spatial data structure will have to be updated every time an object is created or its position is changed. Unfortunately spatial data structure approaches without quadratic complexity in their operations do not make an optimal sort of the space.

The spatial data structure is created and updated while the object model is being updated by the knowledge base. Ideally, the smallest rectangle which enclose each geometry are included in different areas in which the plane is divided. Assuming that consistency between objects exists, the spatial relationships can only change from one to another instant only with the object belonging to the same areas or adjacent areas. With this type of management is very simple and quick to refer the candidate geometries which reduce the number of test in the object model.

Many structures can be used depending on the conditions of each application. Applications with few dynamic objects that are not changing constantly its area may use tree structures. R-Tree [11] or R* [10] approaches could be very proper. In the same way there are also previous and well known structures, like quad trees and k-dimensional trees, these structures are very useful for static object applications. kd-tree approaches can be combined with dynamic objects, however if the

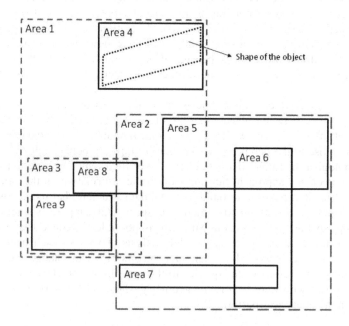

Fig. 3 Example of structured spatial information.

number of dynamic objects grows too this would force to rebuild the tree at every moment and this would not be efficient. If a big quantity of dynamic and static objects is needed, it would be interesting to have another type of structure. [9] is an example of approach with logarithmic complexity search and non-quadratic complexity insertion operations with suboptimal solutions.

4 Case Study: Video Tracking

This type of architecture has multiple areas of application. A direct application area is tracking systems. This type of systems have many needs that this architecture can cover, such as, moving objects over the time, tasks which include spatial and semantic interpretation, good prospects for processing time performance, etc.

Imagine that exist a tracking system which stores all its information on a knowledge base to carry out semantic and spatial tasks for error corrections. These semantic tasks can treat, for example, the problem of loss of consistency in the size of a track when it crosses with another track. The aim of this error correction is to keep the size of the tracks constant during the overlap.

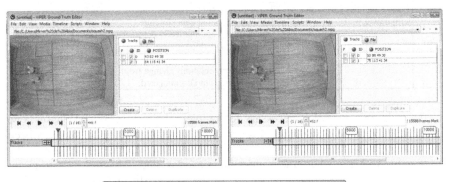

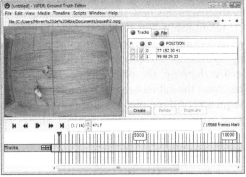

Fig. 4 Error correction in a tracking system. 4A shows the beginning of the error treatment. 4B shows no changes in the size of the tracks during the time of overlap. 4C shows the end of the error treatment when the tracks can change its size.

Table 1 Relationship between events and actions during an example of execution of the system.

Fig	Event	Actions per frame	Topology Relationships
-	Track1 and Track2 are in an approximation state.	1) Track1 and Track2 positions are updated at the spatial data structure. 2) Object model does not detect changes between the geometric relationships.	DC(Track1, Track2)
3A	Track1 and Track2 intersect.	1) Track 1 and Track 2 positions are updated at the spatial data structure. 2) Object model automatically detect the intersection between Track1 and Track2. 3) RCC topology relations change from Disconnected to Externally Connected.	EC(Track1, Track2)
3B	Track1 and Track2 overlap.	1) Track 1 and Track 2 positions are updated at the spatial data structure. 2) Object model automatically detect the overlap between Track1 and Track2. 3) RCC topology relations change from Externally Connected to Partially Overlaps.	PO(Track1, Track2)
3C	Track1 and Track2 are in a withdrawal state.	1) Track 1 and Track 2 positions are updated at the spatial data structure. 2) Object model automatically detect the absence of relationship between Track1 and Track2. 3) RCC topology relations change from Partially Overlaps to Disconnected.	DC(Track1, Track2)

To make this control is necessary to complete the task in several phases; one must first know the positioning of the tracks at time. According to what has been specified, that information must be provided by the knowledge base. Secondly the system has to know when tracks enter in an overlap state. This task is carried out jointly by the two modules belonging to the synergistic approach. The object model automatically discovers this new spatial relationship between the tracks and stores it in as a RCC relationship. Finally the system should keep the size of both fixed during the overlap duration. This condition must be sent from the knowledge base to the tracking system as a recommendation. In the same way, when the tracks are no longer overlapping, the object model has to modify the RCC relationship between them and the submission of recommendations has to be stopped. Figure 3 shows a sequence of three pictures illustrating an example case of automatic error correction with a tracking tool. Table 1 shows the event-action response of the system to the sequence in Figure 4.

5   Conclusion and Future Work

We have presented a novel architecture to address a dual problem; the automatic assertion of dynamic spatial objects and the overhead that these updates may cause on a knowledge base. The architecture has been designed to be embedded in any ontology-based application but some recommendations about possible standards to implement it in a real system have been done.

Future works will include the use of the temporal properties of RCC taking advantage of space operations as operations between intervals. Thus a time interval can be asserted to be included in another one, consecutive, partially coincident, etc. [6] This architecture will be implemented in a semantic activity recognition system similar to that have been proposed in the example application.

Acknowledgments. This work was supported in part by Projects CICYT TIN2008-06742-C02-02/TSI, CICYT TEC2008-06732-C02-02/TEC, CAM CONTEXTS (S2009/ TIC-1485) and DPS2008-07029-C02-02.

References

[1] Randell, D.A., Cui, Z., Cohn, A.G.: A Spatial Logic based on Regions and Connection. In: Proceedings 3rd International Conference on Knowledge Representation and Reasoning (1992)

[2] Grenon, P., Smith, B.: SNAP and SPAN: Prolegomenon to Geodynamic Ontology. IFOMIS Technical Report, University of Leipzig

[3] Grütter, R., Bauer-Messmer, B.: Combining OWL with RCC for Spatioterminological Reasoning on Environmental. In: Proceedings of 3rd OWL: Experiences and Directions, OWLED 2007 (2007)

[4] Grütter, R., Scharrenbach, T., Bauer-Messmer, B.: Improving an RCC-Derived Geospatial Approximation by OWL Axioms. In: Sheth, A.P., Staab, S., Dean, M., Paolucci, M., Maynard, D., Finin, T., Thirunarayan, K. (eds.) ISWC 2008. LNCS, vol. 5318, pp. 293–306. Springer, Heidelberg (2008)

[5] Stocker, M., Sirin, E.P.: A Hybrid RCC-8 and RDF/OWL Reasoning and Query Engine. In: Proceedings of OWL: Experiences and Directions 2009 (OWLED 2009). Sixth International Workshop, vol. 529 (2009)

[6] Gómez-Romero, J., Patricio, M.A., Garcia, J., Molina, J.M.: Towards the Implementation of an Ontology-Based Reasoning System for Visual Information Fusion. Springer, Heidelberg (2009)

[7] Gómez-Romero, J., Patricio, M.A., Garcia, J., Molina, J.M.: Ontology-based context representation and reasoning for object tracking, Colmenarejo (2009)

[8] Schuele, M., Karaenke, P.: Qualitative spatial reasoning with topological information in BDI agents (2010)

[9] Hadjieleftheriou, M., Kollios, G., Tsotras, V.J., Gunopulos, D.: Efficient Indexing of Spatiotemporal Objects. In: Jensen, C.S., Jeffery, K., Pokorný, J., Šaltenis, S., Hwang, J., Böhm, K., Jarke, M. (eds.) EDBT 2002. LNCS, vol. 2287, p. 251. Springer, Heidelberg (2002)

[10] Beckmann, N., Kriegel, H.-P., Schneider, R., Seeger, B.: The R*-tree: An Efficient and Robust Access Method for Points and Rectangles. In: Proceedings ACM SIGMOD International Conference on Management of Data, Atlantic City, pp. 322–331 (1990)

[11] Guttman, A.: R-trees a dynamic mdex structure for spatial searching. In: Proceedings ACM SIGMOD International COnference on Management of Data (1984)

[12] Open Geospatial Consortium,
     http://www.opengeospatial.org/standards

[13] OpenGIS Simple Features Specification for SQL,
     http://www.opengeospatial.org/standards/sfs

A Bayesian Strategy to Enhance the
Performance of Indoor Localization Systems

Josué Iglesias, Ana M. Bernardos, and José R. Casar

Abstract. This work describes the probabilistic modelling of a Bayesian-based mechanism to improve location estimates of an already deployed location system by fusing its outputs with low-cost binary sensors. This mechanism takes advantage of the localization capabilities of different technologies usually present in smart environments deployments. The performance of the proposed algorithm over a real sensor deployment is evaluated using simulated and real experimental data.

Keywords: location management, sensor data fusion, Bayesian inference, pervasive computing, Wireless Sensor Networks.

1 Introduction

The localization problem has received considerable attention in the areas of Ambient Intelligence (AmI) and context-aware services, as many applications need to know where certain objects (or people) are located. Many technologies, such as infrared, ultrasounds and video cameras, may enable solutions (with their own limitation and constraints). In recent years, with the development of Wireless Sensor Networks (WSNs) and wireless communication protocols (WiFi, Bluetooth, ZigBee, UWB, etc.), Radio Frequency-based localization approaches provide another interesting perspective to be considered. Achieving accurate localization using these technologies is a very difficult and often costly operation, being still an open challenge. In these kinds of technological scenarios, one possible alternative is to calculate symbolic locations instead of geographic coordinates; examples of such symbolic locations are room numbers in a building or street addresses in a city.

Moreover, in AmI environments is not unusual to find (direct or indirect) proximity detection mechanisms used, for example, to detect objects usage patterns. These mechanisms are usually supported by a number of binary sensors, those returning a logic '1' if human presence is detected within a certain sensing area,

Josué Iglesias · Ana M. Bernardos · José R. Casar
Telecommunications Engineering School, Technical University of Madrid (UPM), Spain
e-mail: {josue,abernardos, jrcasar}@grpss.ssr.upm.es

J.M. Molina et al. (Eds.): User-Centric Technologies and Applications, AISC 94, pp. 79–92.
springerlink.com   © Springer-Verlag Berlin Heidelberg 2011

otherwise returning a logic '0'.The modality of binary sensors includes sensors such as break-beam, contact, PIR (Passive IntraRed)or binary Doppler-shift sensors; in general, these are low-cost technologies, being currently used in resource-constrained scenarios [1]. The flexibility of WSNs based on programmable nodes allows using them as binary sensors detecting proximity of objects or people. Radio Frequency IDentification (RFID) technology can be also seen as a binary sensor technology.

WSN and non-expensive binary sensors opens up new application fields, introducing a new paradigm of sensing where, instead of having a few set of expensive sensors with advanced functions and high performance, many sensors with diverse functions and performance can be distributed in order to collaborate to enhance location estimation. Data fusion techniques can combine data gathered from multiple sensing sources. These techniques include e.g. Bayesian and Dempster-Shafer inference, aggregation functions, interval combination functions or classification methods [2].

In particular, Bayesian techniques are used to manage data coming from unreliable sources, as they cope with uncertainty and sensor errors. Bayesian methods offer a formal way to update existing knowledge from new evidences, providing an adaptive learning tool that offers more accurate estimates as the number of processed measurements increases. Bayesian techniques can be used to develop adaptive systems that respond to real-time data input and improve their performance over time.

This paper proposes to apply a Bayesian algorithm to fuse information from a symbolic localization system working over a ZigBee WSN with data from binary sensors (short-range motes and RFID sensors), in order to enhance the accuracy and stability of the location estimates. Simulated results, built on different distributions of sensors offering variable sensing quality, show the feasibility of the proposed approach.

Next section surveys relevant related work concerning localization estimation improvement with probabilistic fusion techniques. Section 3 exemplifies the type of technologies and infrastructure on top of which the algorithm works with the deployment that has been built in our lab for testing purposes. The Bayesian algorithm used is detailed in Section 4; simulated experiments and results are gathered in Section 5. Finally, Section 6 discusses on the obtained data and presents our ongoing works.

2 Related Works

Although other approaches exist, RF-based location technologies often process received signal strength measurements (RSS) to determine the position of a target moving around in the coverage area of an infrastructure composed of several anchor nodes. Channel modelling, fingerprinting and proximity (also known as cell-identification) techniques are the main methods working on RSS. Model-based techniques use a channel model to describe the propagation between the mobile target and the anchor nodes, providing a direct relationship between the RSS and the distance between a pairs of nodes ([3, 4, 5]). Fingerprinting-based methods consist of creating a 'radio map' with the observed RSS from different anchor nodes at different positions ([5, 6]); this off-line calibrated map is used during the on-line localization phase to identify the spatial point(s)which minimizes a defined 'distance' function between the on-line RSS sample and the stored fingerprint. The main limitation of signal-strength based techniques is that RSS has a fluctuating nature, especially when indoors, due to propagation effects like shadowing, multipath, etc. This is why achieving accurate localization based on RSS is still an open challenge, being the accuracy of these kinds of methods only around 2 or 3 meters [7]. Proximity-based approaches [8, 9, 10, 11] estimate locations from connectivity or visibility information. In WSNs, mobile devices are typically connected to the anchor node they are closest to, so the location of the node can be estimated as the same as the location of the anchor node to which it is connected [9, 10]. In [8, 11], the unknown location is estimated using connectivity information to several nodes. The strength of 'cell-ID' techniques is their fairly simple implementation and modest HW requirements. Nevertheless, only a very coarse-grained location can be estimated, requiring a very dense grid of fixed nodes in order to reach small granularities [12].

These indoor location systems are usually providing a horizontal functionality enabling different types of vertical services. These services may have different needs for location accuracy and granularity [3], then being also service-dependant the decision about which localization system to use. Apart from traditional RF technologies (such as WiFi or Bluetooth), RFID may be used to support localization. For example, inexpensive passive RFID sensors may work as 'proximity' sensors, delivering information about whether the user is close to a certain position or not (due to the short-range –cms.– of the technology).A review of active and passive RFID-based location algorithms can be found in [13] (including multilateration, Bayesian inference, nearest neighbour, proximity and kernel-based learning methods). From a general perspective, ideal, real (imperfect) and directive binary proximity sensors are deeply modelled in [14, 15,16], respectively. Although in these cases binary sensors are used for target detection to offer tracking services, these models could be applied when proposing strategies to improve localization estimation.

Data fusion techniques may be used to integrate the information obtained from different sensor sources ([1, 2, 17] for information fusion surveys in WSNs).Fox et al. survey Bayesian filtering techniques for multi sensor fusion [1], arguing that probabilistic fusion methods are heavy in terms of computational load, requiring a centralized infrastructure to run the algorithms. A symbolic wireless localization device using a Bayesian network to infer the location of objects covered by IEEE 802.11 wireless network is developed in [18], where RSS signal-to-noise ratios received from different access points are quantified (also modelling a variable 'noise' that could affect the measures). Simple binary sensors in a Bayesian framework are also used in [19] to provide room-level location estimation and rudimentary activity recognition. In [20] a HMM (Hidden Markov Model) is used to stabilize a Bayesian-based location inference output in a WiFi-based localization system. In the domain of mobile robotics, RFID and Bayesian inference is used to perform obstacle detection, mitigating multipath effects [21, 22].

A recursive Bayesian estimator, integrating WSN-based location data and kine-matic information, is presented in [23].

Most of the proposals above employ Bayesian inference concepts in the design stage of the localization system and infrastructure. Moreover, they work on a single technology, not combining different ones. Our approach makes possible to improve a previously existing localization system by adding low-cost sensors, and takes advantage of the localization capabilities of different technologies usually present in smart environments.

3 Problem Statement

The particular sensing characteristics of the deployment where our improvement mechanism has been applied are probabilistically described in this section.

3.1 Description of the Technological Scenario

The Bayesian algorithm described in this paper is thought to work on a previously deployed localization system providing symbolic (zone-based) localization. This system is characterized by its accuracy, modelled as the error rate that determines the percentage of successful estimates it provides in every potential localization scenario.

In practice, in our laboratory we have deployed a WSN (ZigBee) localization infrastructure composed by twelve anchor nodes (MicaZ motes), two for each of the rooms available in our particular deployment (Figure1). A simple proximity-based location algorithm is used to estimate the zone in which a mobile node is at a time:$o(t) = \{0,1, ..., N_Z - 1\}$. In brief, the mobile node is assumed to be in the zone where the anchor node emitting the strongest detected RSS is located [9, 10].

As previously stated, indoor environments such as the one described here are electromagnetically unstable due to the continuous 'reconfiguration' of obstacles causing multipath and fading (people randomly moving around, periodic refurbishment and re-collocation of objects, people's clustering, etc.) and also due to humidity and temperature variability. This situation provokes adverse effects in RSS-based localization systems' accuracy. For example, our simple localization system is capable of locating a mobile target with an average error of 28.79%; this error varies between 3.4% and 75.62% depending on the area of work.

In these circumstances, the performance of the localization system may be modelled as the probability for the mobile target of being detected in $o(t)$ when actually being in $H_k(t)$; that's to say: $P(o(t)|H_k(t))$, being $H_k(t) = \{0,1, ..., N_Z - 1\}$ a variable representing the actual mobile target's location over time.

Then, our objective is to enhance the localization system's accuracy with a fusion strategy that will integrate: 1) additional information about the deployment facilities(i.e., possible and impossible transitions between zones),2) proximity detection to reference points and 3) transition detections between zones. Both

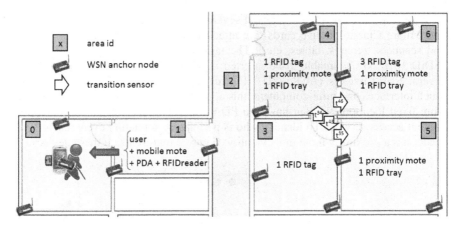

Fig. 1 Technological scenario in our laboratory(adapted from [24]).

proximity and transitions will be detected by using binary sensors, as explained in the next subsection.

In our prototype, users will need to carry a mobile mote to be detected as part of the WSN. They will also carry a RFID reader attached to their PDAs that will also identify them when interacting with any RFID-tagged location or object. As the user can be easily identified, our approach could support multiusers without being redesigned.

3.2 A Probabilistic Model for Proximity Binary Sensors

Proximity binary sensors are those that output a '1' when the target of interest is within its sensing range and '0' otherwise. Three types of proximity binary sensors have been selected to be deployed in the above deployment: passive RFID sensors, pressure mats and Zigbee motes(which emitting power has been tuned to reduce their coverage area). Pressure mats and passive RFID provide information about the user's proximity with high accuracy, due to their very short range, while low-coverage motes will serve to accurately detect presence around established points of interest. Of course, other type of sensors (even logical ones – those detecting computers or appliances' connectivity) and technologies may be additionally considered as binary detectors of proximity in smart environments.

In practice, in our lab deployment, the basic proximity binary sensor infrastructure is composed of two motes: an anchor node, to be settled in the zone where proximity needs to be detected, and a mobile node configured to work at minimum power (-25dBm.) in order to restrict its coverage area to, approximately, 1.5 meters. The mobile node is constantly sending proximity beacons that are received by those in the infrastructure when in their coverage area.

Pressure mats are placed on chairs, sofas, carpets, waypoints, etc., in order to detect users sitting or standing on them. Each pressure mat has a mote attached, being able to transmit the sensed data through the WSN.

With respect to RFID, on one hand several passive tags (Mifare Ultralight tags and Mifare Classic passive cards) are attached to objects (access doors, printers and scanners, screens, tables, etc.). The tags may be read using a RFID reader (SDiD 1010 NFC) suitable to be inserted in the user's PDA. On the other hand, several RFID readers (ACR122U NFC readers) have been connected to some touch interfaces and mini-computers; this approach enables reading tagged objects from fixed locations without having a PDA, being useful for example to support content access services.To identify who is performing a reading event, each RFID reader has a proximity mote in its vicinity. Figure 2 represents all these devices.

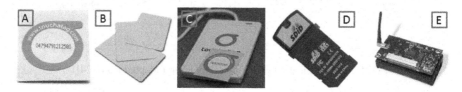

Fig. 2 Binary proximity sensors used in our lab deployment: A), B) passive RFID tags, C) RFID reader with USB interface, D) embedded RFID reader and E) MicaZ mote.

Probabilistically, the set of proximity binary sensors estimations are defined by $\overline{D(t)} = \left[\overline{D^M(t)}, \overline{D^P(t)}, \overline{D^R(t)}\right]$, using M index for power-tuned motes, P for pressure mats and R for RFID-based sensors. Being N_D^x the total number of proximity sensors of type $x = \{M, P, R\}$, then $\overline{D^x(t)} = \left[d_0^x(t), d_1^x(t), \ldots, d_{N_D^x-1}^x(t)\right]$ and $d_n^x(t) = \{0,1\} \; \forall 0 < n < N_D^x - 1$.

Proximity binary sensors' detection events are modelled as $P(d_n^x(t)|c_n(t))$, being $c_n(t) = \{0,1\}$ the variable that states if the mobile target (the user) is actually within the sensor range.

3.3 A Probabilistic Model for Transition Binary Sensors

Transition binary sensors detect user's transfer between adjacent zones (returning '0' when no transition occurs and '1' otherwise). Although other practical approaches are possible, in our deployment the transition binary sensor is a virtual sensor which process data coming from a pair of pressure mats, each of them placed on both sides of selected waypoints (not every waypoint has a transition sensor installed). Transition sensors estimate the direction of the movement depending on the sequence of activation of the pressure mats. Each pair of pressure mats has a mote attached for communication purposes.

The set of transition binary sensors estimations are defined by $\overline{I(t)} = \left[i_0(t), i_1(t), \ldots, i_{N_I-1}(t)\right]$, being N_I the number of transition sensors deployed and $i_n(t) = \{0,1\} \; \forall 0 < n < N_I - 1$.

Transition binary events are probabilistically modelled as $P(i_n(t)|r_{pq}(t))$, being $r_{pq}(t) = \{0,1\}$ the variable that states if the mobile target (the user) actually transits or not between the areas where sensor i_n is deployed (i.e., between areas p and q).

Future extensions of the deployment may consider using contact sensors (monitoring the use of doors, cabinets, etc.) and/or sets of motes configured as transition binary sensors.

4   A Bayesian Strategy to Enhance an Indoor RSS Zone-Based Localization System

Once modelled our system inputs (location system and binary sensors), this section formalizes the probabilistic relationships between them (4.1) and describes the inference algorithm used to exploit them (4.2).

4.1   Stage 1: Formulating the Problem through a Dynamic Bayesian Network

The dynamic Bayesian network (DBN) structure in Figure 3 is proposed to establish relationships between sensor events coming from the three sources of localization information: $o(t)$, $\overline{D(t)}$ and $\overline{I(t)}$ (localization events, proximity events and transition events), which are from now on handled as random variables. DBNs characteristics perfectly fit in the fusion problem here described as they model systems that are dynamically changing or evolving over time, enabling to monitor and update the system as time proceeds (and even predict further behaviours). Accordingly to this DBN structure, each sensor estimation is considered probabilistically independent from any other estimation.

It has to be reminded that random variable $H_k(t)(k = \{0,1,\ldots,N_Z - 1\})$ is defined to represent the real user location over time.

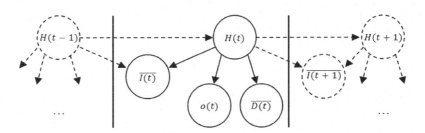

Fig. 3 Dynamic Bayesian network structure

4.2   Stage 2: Applying Bayesian Estimation to Enhance Localization Estimates

Sensors estimations are fused using a recursive Bayesian filter (RBF). The RBF is a general probabilistic technique to recursively estimate an unknown state using incoming measurements [25]. In our particular problem, $H(t)$ is the unknown state and $o(t)$, $\overline{D(t)}$ and $\overline{I(t)}$ are the incoming measurements.

Being $\overline{z(t)} = \left[o(t), \overline{D(t)}, \overline{I(t)}\right]$ and $\overline{\mathbb{Z}(t)} = \left[\overline{z(t)}, \overline{z(t-1)}, \dots, \overline{z(1)}\right]$ respectively the whole set of measurements in t and the whole set of measurements since the system is started, the improved location estimation ($\widehat{H(t)}$) can be determined by calculating the area (argument k) that maximizes the probability of being there when taking into account the sensing history:

$$\widehat{H(t)} = arg \max_{k}\left[P(H_k(t)|\overline{\mathbb{Z}(t)})\right] \tag{1}$$

In order to be able to develop (1), it is necessary to consider that:

- temporal transitions between states follows a Markovian evolution;
- for localization and proximity events, the measurements observed at a time are independent from previous states (see causal relations in Figure3's DBN);
- for transition sensors, the measurements observed at a time depend on the state in that particular instant but on the previous instant too (see causal relations in Figure3's DBN).

Then, having into account that:

$$P(H_k(t)|\overline{\mathbb{Z}(t)}) = P(H_k(t)|\overline{\mathbb{Z}(t-1)}, \overline{z(t)}) \tag{2}$$

and applying Bayes theorem to (2):

$$P(H_k(t)|\overline{\mathbb{Z}(t)}) = \frac{P(\overline{z(t)}|\overline{\mathbb{Z}(t-1)}, H_k(t)) \cdot P(H_k(t)|\overline{\mathbb{Z}(t-1)})}{P(\overline{z(t)}|\overline{\mathbb{Z}(t-1)})} \tag{3}$$

The first term in the numerator of equation (3) corresponds to the sensors' event models, existing some differences between proximity and transition sensors:

- for proximity binary sensors, as they are independent over time:

$$P\left(\overline{z(t)}\middle|\overline{\mathbb{Z}(t-1)}, H_k(t)\right) = P\left(\overline{z(t)}\middle|H_k(t)\right) \tag{3.1a}$$

- for transition binary sensors:

$$P\left(\overline{z(t)}\middle|\overline{\mathbb{Z}(t-1)}, H_k(t)\right) = \sum_{\forall l}\left[P\left(\overline{z(t)}\middle|H_k(t), H_l(t-1)\right) \cdot P(H_l(t-1)|\overline{\mathbb{Z}(t-1)})\right] \tag{3.1b}$$

According to the DBN structure, the second term in equation (3) can be also developed as:

$$P(H_k(t)|\overline{\mathbb{Z}(t-1)}) = \sum_{\forall l}\left[P(H_k(t)|H_l(t-1)) \cdot P(H_l(t-1)|\overline{\mathbb{Z}(t-1)})\right] \tag{3.2}$$

Summarizing, the RBF algorithm requires defining 1) a probabilistic model for every kind of sensor and 2) a probabilistic user transition model representing possible movements between zones. Following, the implications of this need are explained.

4.3 Sensor Model

RBF algorithm requires having the sensors modelled as stated in equations (3.1). Regarding proximity sensors, equation (3.1a) probabilistically relates the sensor estimation $(z(t) = d_n^x(t) = \{0,1\})$ to the real area where the user is placed $(H_k(t))$. However, the proximity binary sensors are modelled as defined in Section 3.2. These two expressions can be related as follows:

$$P\big(d_n^x(t)\big|H_k(t)\big) = \sum_{\forall s = \{0,1\}} \big[P(d_n^x(t)|c_n(t) = s) \cdot P\big(c_n(t) = s\big|H_k(t)\big)\big] \quad (4)$$

So it is necessary to model $P(c_n(t)|H_k(t))$ in order to be able to apply RBF algorithm. That means that we need to know certain *a priori* user behaviours patterns (i.e., how often is the user in the coverage of each sensor, knowing that he/she is in a particular area). In this very first implementation, this issue is approached having into account the sensors' coverage, so $P(c_n(t)|H_k(t)) = areaCoverage_{sensor^n}/areaZone^k$. Each sensor is also supposed to just have coverage in only one area.

For the localization system and the transition sensors the same reasoning may be applied. In these cases quality models presented in Sections 3.1 and 3.3 perfectly fits in the RBF algorithm, not requiring extra information about user behavioural patterns.

4.4 Transition Model

RBF algorithm also requires having a transition model (determined by $P(H_k(t)|H_l(t-1))$). As a simple approach to model the transition, our transition model now only states impossible transitions, depending on the physical map of the environment, and equidistributes the possible ones (i.e., the probability of staying in a particular area is the same as transiting to any of the areas with direct access).

5 Simulation Results

This section gathers some simulations that illustrate the algorithm's performance. With respect to sensors' event modelling, we have configured experimental models from the real detection error rates both for proximity sensors (only those based on motes) and for the localization system. However, transition sensors' quality has not been empirically estimated yet, so its influence over the algorithm output has been evaluated for different ranges of qualities.

5.1 Analysing the Effects of the Sensor Deployment Strategies

In order to roughly determine the overall capabilities of the improvement algorithm some preliminary simulation tests have been run. Table 1 shows the simulation parameters and Figure4 the results. Areas distribution, location system and
proximity sensors characteristic have been modelled according real laboratory
conditions and measurements; however, for these simulations, location of each
proximity and transition sensors is randomly settled. This process is repeated several times in order to obtain average results.

These preliminary tests will help in evaluating how the number of proximity
and transition sensors and the features (quality) of the transition sensors may affect the location improvement algorithm output.

Table 1 Preliminary simulation tests parameters

Transition model	$\begin{cases} P(H_k(t)\|H_l(t-1)) = 0 & \textit{if } k \textit{ and } l \textit{ are not communicated} \\ \textit{equidistrubuted} & \textit{otherwise} \end{cases}$
Location system quality	Real data from a WSN-based location system (proximity-based) Experimental average hit rate = 71.21%
#proximity sensors	From 1 to 46 (1 sensor → only 1 area of coverage) Randomly allocated (in each scenario)
Proximity sensors quality	Every sensor modelled as low powered motes proximity sensors Experimental sensor quality: $\begin{cases} P(d_n^x(t) = 0\|c_n(t) = 0) = 1 \\ P(d_n^x(t) = 1\|c_n(t) = 1) = 0.978926 \end{cases}$
#transition sensors	From 1 to 12 (maximum according real area distribution, Figure1) Randomly allocated (in each scenario)
Transition sensors quality	Ranging from 90 to 100% in positive hit rate [100% in negative hit rate]
#random scenarios = 1,000	#trajectories per scenario = 1,000 #transitions per trajectory = 1,000

Figure 4.a depicts the hit rate depending on the number of proximity sensors,
showing also the introduced improvement with respect to the location system
quality. The improvement introduced by the Bayesian algorithm reaches 99.0,
99.5 and 99.9 hit rate % with 48, 58and 87proximity sensors respectively. Considering the real scenario presented in Figure1, using 11 proximity sensors would
lead to 89.96% of hit rate and an improvement of +18.75% with respect to the
original location system quality.

It can be noticed that with no sensors the improvement algorithm obtains
around 78.35% of hits; this improvement (~ +7%) is due to the transition model
effects that corrects impossible transitions estimated by the location system.

Hit rate and improvement are also shown in the second figure, now depending
in the number and quality of the transition sensors. Figure 4.b also confirms that
the more transition sensors and the more quality they have, the more improvement
is achieved in the localization estimation. Anyway, this test also reveals a lower
threshold regarding the quality of transition sensors: no significant improvements
are achieved employing transition sensors of lower quality than around 85%. This
issue must be taken into account when developing real binary transition sensors.

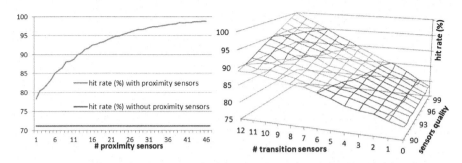

Fig. 4 a) Proximity and b) transition sensors influence over location estimation

5.2 Evaluation of the Bayesian Location Improvement Algorithm in a Real Scenario

Finally, in Figure 5 the results of a simulation employing together both proximity and transition sensors are presented. This test has been set to be run using our real deployment configuration (Figure1). The number of proximity sensors has been set to 11 for this simulation (matching the number of sensors nowadays available for our real deployment). Only 4 transition sensors have been employed, placed in the 'crossroads' of areas 3, 4, 5 and 6. These four areas share a common transition zone, so it was decided to place there several transition sensors, trying to reduce location system errors. As we are still working in the configuration of the transition sensors, the obtained improvement is shown over several transition sensors qualities.

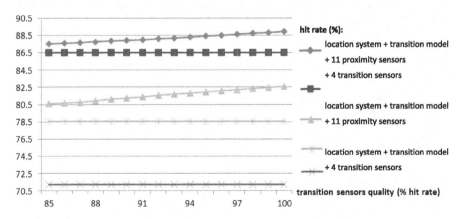

Fig. 5 Real deployment influence over location estimation

When only using 11 proximity sensors to enhance the location system (no transition sensors), an improvement of +15.29% is obtained (86.5% of global hit rate) being these results slightly lower than in the previous test (Figure4.a). The fact that in the tests of previous subsection sensors are randomly placed over any of the areas may have influenced these results. It has to be noticed that in Figure 5 results, all the 11 proximity sensors are placed in only four of the total seven areas available. The same effect applies to the transition sensors.

6 Discussion and Further Work

In this paper we have proposed a probabilistic Bayesian-based model to be applied to an already deployed location system in order to enhance its quality by means of adding new information coming from low-cost binary sensors, thus taking advantage of the localization capabilities of different technologies usually present in smart environments (programmable mote-based WSN nodes, RFID tags and readers, etc.). The set of sensors used (location system and proximity / transition binary sensors) and its relations are also probabilistically modelled. The performance of the proposed algorithm over a real sensor deployment is evaluated simulating the enhancement mechanism using real experimental data (specifically regarding location system and proximity sensors events model).

Several simulations have been performed, taking into account random binary sensors deployment but also a fixed one based on the particular resources available in our laboratory. For the fixed deployment, these simulations offer improvements in location hit rate ranging from around +16.3% and +17.7% (depending on specific sensors qualities), increasing location system quality from a 71.21% of hit rate to around 88%.These simulations are also a good starting point for quantifying and qualifying future real implementations of this improvement mechanism. In this sense, simulations performed in a fixed deployment should be extended in order to further analyse the influence of particular placements of binary sensors. The results regarding transition sensors have also to be taken into account when actually configuring these kinds of sensors.

The improvement mechanism also can be further enhanced. For example, currently it is assumed that the location system generates a deterministic estimate but it may be generalized to allow the use of a probability distribution.

We are working on a real implementation of this mechanism, which would require 1) to actually configure a set of transition binary sensors, 2) to probabilistically model different transition and proximity sensors (i.e., RFID-based) and 3) to design the infrastructure processing the data coming from the different sensors.

Having all these issues in mind, a real implementation of these improvement mechanisms appears to be quite promising.

Acknowledgments. This work has been supported by the Government of Madrid under grant S2009/TIC-1485 (CONTEXTS) and by the Spanish Ministry of Science and Innovation under grant TIN2008-06742-C02-01.

References

1. Teixeira, T., Dublon, G., Savvides, A.: A Survey of Human Sensing: Methods for Detecting Presence, Count, Location, Track and Identity. ACM Computing Surveys (2010)

2. Nakamura, E.F., Loureiro, A.A.F., Frery, A.C.: Information fusion for wireless sensor networks. ACM Computing Surveys 39(3), 9-es (2007)
3. Li, X.: RSS-based location estimation with unknown pathloss model. IEEE Transactions on Wireless Communications 5(12), 3626–3633 (2006)
4. MacDonald, J.T., Roberson, D.A., Ucci, D.R.: Location estimation of isotropic transmitters in wireless sensor networks. In: Military Communications Conference, pp. 1–5 (October 2006)
5. Bahl, P., Padmanabhan, V.N.: RADAR: an in-building RF-based user location and tracking system. In: Proc. IEEE Infocom, pp. 775–784 (March 2000)
6. Lorincz, K., Welsh, M.: 'Motetrack: a robust, decentralized approach to RF-based location tracking. In: Int. Workshop on Location and Context-Awareness at Pervasive (May 2005)
7. Martín, H., Tarrío, P., Bernardos, A.M., Casar, J.R.: Experimental Evaluation of Channel Modelling and Fingerprinting Localization Techniques for Sensor Networks. In: International Symposium on Distributed Computing and Artificial Intelligence 2008 (DCAI 2008), vol. 50, pp. 748–756. Springer, Berlin (2009)
8. Bahl, P., Padmanabhan, V.N.: Radar: An in-building rf-based user location and tracking system. In: INFOCOM (2000)
9. Giorgetti, G., Gupta, S.K., Manes, G.: Wireless localization using self-organising maps. In: IPSN 2007 (2007)
10. He, T., Huang, C., Bium, B.M., Stankovic, J.A., Abdelzaher, T.: Range-free localization schemes for large scale sensor networks. In: ACM MobiCom 2003 (2003)
11. Wark, T., Crossman, C., Hu, W., Guo, Y., Valencia, P., Sikka, P., Corke, P., Lee, C., Henshall, J., O'Grady, J., Reed, M., Fisher, A.: The design and evaluation of a mobile sensor actuator network for autonomous animal control. In: IPSN, pp. 206–215 (2007)
12. Kaseva, V.A., Kohvakka, M., Kuorilehto, M., Hännikäinen, M., Hämäläinen, T.D.: A Wireless Sensor Network for RF-Based Indoor Localization. EURASIP Journal on Advances in Signal Processing 27, Article ID 731835 (2008), doi:10.1155/2008/731835
13. Zhou, J., Shi, J.: RFID localization algorithms and applications—a review. Journal of Intelligent Manufacturing 20(6), 695–707 (2008)
14. Wang, Z., Bulut, E., Szymanski, B.K.: A distributed cooperative target tracking with binary sensor networks. In: Proc. IEEE International Conference on Communication (ICC) Workshops, Beijing, China, May 2008, pp. 306–310 (2008)
15. Wang, Z., Bulut, E., Szymanski, B.K.: Distributed Target Tracking with Imperfect Binary Sensor Networks. In: Proc. IEEE Global Communications Conference (Globecom), New Orleans, USA, December 2008, pp. 1–5 (2008)
16. Wang, Z., Bulut, E., Szymanski, B.: Distributed Target Tracking with Directional Binary Sensor Networks. In: Global Telecommunications Conferece. GLOBECOM (2009)
17. Wymeersch, H., Lien, J., Win, M.: Cooperative Localization in Wireless Networks. Proceedings of the IEEE 97(2), 427–450 (2009)
18. Castro, P., Chiu, P., Kremenek, T., Muntz, R.: A probabilistic room location service for wireless networked environments. In: Abowd, G.D., Brumitt, B., Shafer, S. (eds.) UbiComp 2001. LNCS, vol. 2201, p. 18. Springer, Heidelberg (2001)
19. Wilson, D.H.: People Tracking Using Anonymous, Binary Sensors. M.S. Thesis, CMU Center for Automated Learning & Discovery (2003)

20. Ladd, A., Bekris, K., Rudys, A., Wallach, D., Kavraki, L.: On the feasibility of using wireless ethernet for indoor localization. IEEE Transactions on Robotics and Automation 20(3), 555–559 (2004)
21. Joho, D., Plagemann, C., Burgard, W.: Modeling RFID signal strength and tag detection for localization and mapping. In: IEEE International Conference on Robotics and Automation, 2009, pp. 3160–3165 (2009)
22. Jia, S., Sheng, J., Takase, K.: Improvement of performance of localization ID tag using multi-antenna RFID system. In: SICE Annual Conference, pp. 1715–1718 (2008)
23. Klingbeil, L., Wark, T.: A Wireless Sensor Network for Real-Time Indoor Localisation and Motion Monitoring. In: Klingbeil, L. (ed.) 2008 International Conference on Information Processing in Sensor Networks (IPSN 2008), pp. 39–50 (2008)
24. Martin, H., Metola, E., Bergesio, L., Bernardos, A.M., Iglesias, J., Casar, J.R.: An RFID-enabled framework to support Ambient Home Care Services. In: Third International EURASIP Workshop on RFID Technology, Cartagena, Spain (2010)
25. Thrun, S., Burgard, W., Fox, D.: Probabilistic Robotics. MITPress, Cambridge (2005)

A Low Power Routing Algorithm for Localization in IEEE 802.15.4 Networks

Luca Bergesio, Paula Tarrío, Ana M. Bernardos, and José R. Casar

Abstract. Many context-aware applications rely on the knowledge of the position of the user and the surrounding objects to provide advanced, personalized and real-time services. In wide-area deployments, a routing protocol is needed to collect the location information from distant nodes. In this paper, we propose a new source-initiated (on demand) routing protocol for location-aware applications in IEEE 802.15.4 wireless sensor networks. This protocol uses a low power MAC layer to maximize the lifetime of the network while maintaining the communication delay to a low value. Its performance is assessed through experimental tests that show a good trade-off between power consumption and time delay in the localization of a mobile device.

Keywords: Routing, wireless sensor networks, low power listening, localization.

1 Introduction

User's location is a valuable source of information in many applications, including augmented reality, location-based services or interactive user experience [1-2].A common approach to determine the position of a user in an outdoor environment is to use a GPS device. However, GPS is not so effective in indoor environments, and other technologies, such as Wi-Fi, Bluetooth or ZigBee can be used instead. In this paper we consider ZigBee and, in particular, the lower layers of its communication stack (IEEE 802.15.4 [3]). This technology was designed for Low-Rate Wireless Personal Area Networks (LR-WPANs) and has become very popular for wireless sensor networks (WSN). It is characterized by low data rate, low radiofrequency (RF) power, long battery life and low cost of the devices.

A typical location-aware IEEE 802.15.4 network is composed of several static nodes with known positions and one or several mobile nodes whose positions need to be calculated at different time instants. In order to estimate a mobile node's position, the static nodes measure the RSS of the messages sent by the mobile

Luca Bergesio · Paula Tarrío · Ana M. Bernardos · José R. Casar
Data Processing and Simulation Group, ETSI. Telecomunicación
Universidad Politécnica de Madrid, Madrid, Spain
e-mail: {luca.bergesio,paula,abernardos,jramon}@grpss.ssr.upm.es

J.M. Molina et al. (Eds.): User-Centric Technologies and Applications, AISC 94, pp. 93–102.
springerlink.com   © Springer-Verlag Berlin Heidelberg 2011

node (or vice versa). From a set of RSS measurements the position of the mobile node can be calculated using a localization method based on fingerprinting [4] or channel modeling [5]. Due to the limited transmission range of IEEE 802.15.4 devices, covering a large deployment area requires the use of a multi-hop routing algorithm to collect the RSS data in a central server or to disseminate location information over the network.

In this paper we consider the problem of providing a routing mechanism to perform centralized RSS-based localization in an IEEE 802.15.4 network covering a large deployment area. Since the nodes may be battery powered, our main objective is to minimize power consumption and maximize the lifetime of the network. We also consider the time delay in getting the location information as another important aspect. The structure of the rest of the paper is as follows. Section 2 provides some background on energy saving techniques for wireless networks. Section 3 describes a method to perform the RSS data collection using a low power multi-hop routing algorithm, which is evaluated in Section 4. Section 5 presents some conclusions and future research work.

2 Background

One of the most common and effective techniques to reduce power consumption in wireless networks consists in switching off the radio transceiver as much as possible, because it is the most energy consuming hardware component of a node [6]. Since the RF transceiver is usually controlled by the MAC layer, numerous MAC protocols have been proposed in the literature to reduce energy consumption.

Some MAC protocols use time slots to synchronize the communications between the nodes (e.g. S-MAC, T-MAC, etc.) [7]. These algorithms are quite complex and require precise timers, a resource that is not usually available in WSN devices. Another class of MAC protocols is based on CSMA-CA: they do not require synchronization, but they listen to the channel before transmitting, and only transmit if it is clear. Since this approach does not require synchronization, it is quite easy to implement on simple devices.

TinyOS [8],an open source operating system developed by the University of California at Berkeley and widely used in WSN research, supports some not-synchronized protocols, namely, Berkeley MAC (BMAC) [9] and low power listening (LPL) [10]. BMAC is a beta version and will be probably replaced by XMAC [11] on the next version of TinyOS. LPL is stable and available on TinyOS 1.x and 2.x.

In the contributing code to TinyOS 2.x repository [12], there are some examples of routing protocols based on low power listening MAC. One of those is the Ad hoc On-Demand Distance Vector (AODV) routing protocol [13], one of the most popular algorithms for WSNs. This algorithm finds the shortest routes between the nodes and the base station, but leaves the management of acknowledgements and retransmissions to higher levels of the communication stack. The routing algorithm proposed in this paper is also based on a low-power MAC, but tries to find the quickest route from the base station to the other nodes, in order to introduce shorter delays in the communications. Furthermore, it handles acknowledgments and retransmissions in order to guarantee the reception of the RSS measurements collected by the nodes.

3 Proposed Algorithm

In this section we describe a routing mechanism specifically designed to collect RSS measurements in an IEEE 802.15.4 network covering a large deployment area. The proposed algorithm aims at collecting the measurements that are required to perform the localization of a node using multi-hop communications (to cover the whole deployment area) and low-power mechanisms (to extend the lifetime of the network as much as possible).

Our routing algorithm is based on the LPL MAC layer of TinyOS and *opportunistic routing*[14]. The opportunistic routing is based on the availability of broadcast transmissions in wireless networks: a node can communicate simultaneously with all the nodes within its communication range.

Using the idea of opportunistic routing and the LPL MAC protocol, our algorithm traces the paths between each node and the root on an on-demand basis. Then, the nodes use these paths to report the RSS measurements to the base station. We next describe the MAC and routing layers of the proposed algorithm.

3.1 MAC Layer

Our routing algorithm works on top of the LPL MAC algorithm offered by TinyOS, which does not need synchronization and is quite easy to configure and control. The LPL algorithm uses a duty cycle mechanism to put the node into *sleep mode* periodically and save energy. The only parameter that needs to be set is the duration of a cycle, which determines the duty cycle (active listening time cannot be configured and it is set by default to the minimum value permitted by the RF transceiver). With LPL each device listens to the channel periodically and go back to sleep.

If a node wants to transmit a message to another node, it must transmit during a full cycle to ensure that the receiver is listening to the channel while the packet is being sent.

In TinyOS this long transmission is made by sending many small packets covering the whole cycle. Each of these packets may contain either useful data or simply preamble information. In our implementation we have chosen to send these packets with useful information (data packets).

3.2 Routing Protocol

The routing protocol consists of two phases. The first one is the *discovery phase*, in which the routing algorithm finds the routes between the nodes of the network (in particular, the routes to reach the base station from any other node). These routes are then used in the second phase, the *reply phase*, when a node needs to send data packets to another node (in particular, when a node sends the collected RSS information back to the base station).

3.2.1 Discovery Phase

We assume that, at the beginning of the first phase, the nodes do not have any information about which are their neighbors or about the position of the mobile node. Under these circumstances, the easiest way to reach the mobile node is to broadcast a *start transmission* message from the base station. Then, each node retransmits the message once, generating a flooding. With this technique we ensure that each node receives the message, including the mobile node.

This *start transmission* message carries six fields: source address, destination address, sequence number, intermediate source, hop count and action, as shown in Fig. 1. The first is the address of the base station, typically zero. The second is the address of the mobile node we want to localize. It is possible to handle various mobile devices by changing this address. The sequence number is used to identify the packet, and thus, to avoid duplicate deliveries during the flooding mechanism. The intermediate source address is the most important field for the routing algorithm. It stores the address of the node that has most recently retransmitted the message. This address will represent the next hop towards the base station during the reply phase and it is saved in the node's routing table. The hop count stores the number of hops; it is used only for debugging and statistic purposes. The last field is the command we want to send to the mobile node and can be set either to *start* or to *stop* (we will consider other possibilities in the future).

```
 0                   1                   2                   3
 0 1 2 3 4 5 6 7 8 9 0 1 2 3 4 5 6 7 8 9 0 1 2 3 4 5 6 7 8 9 0 1
+-+-+-+-+-+-+-+-+-+-+-+-+-+-+-+-+-+-+-+-+-+-+-+-+-+-+-+-+-+-+-+-+
|          Source Address         |       Destination Address       |
+-+-+-+-+-+-+-+-+-+-+-+-+-+-+-+-+-+-+-+-+-+-+-+-+-+-+-+-+-+-+-+-+
|         Sequence Number         |  Intermediate Source Address   |
+-+-+-+-+-+-+-+-+-+-+-+-+-+-+-+-+-+-+-+-+-+-+-+-+-+-+-+-+-+-+-+-+
|           Hop Count             |            Action              |
+-+-+-+-+-+-+-+-+-+-+-+-+-+-+-+-+-+-+-+-+-+-+-+-+-+-+-+-+-+-+-+-+
```

Fig. 1 Structure of a *start transmission* message

When the mobile node receives the *start* command it begins to transmit packets periodically (it stops transmitting when it receives the *stop* command). When this occurs, some of the static nodes (those that are close to the mobile node) receive these packets and measure their RSS. The next step consists in sending these RSS values to the base station.

3.2.2 Reply Phase

At this point each static node knows the address of a neighbor towards the base station (the intermediate node memorized in the discovery phase). The static nodes that have measured the RSS from the packets sent by the mobile node will retransmit this information to their next-hop neighbors, and the procedure will be repeated until all the RSS measurements reach the base station.

Clearly, if a link fails at a given moment, the packet will not be received by the base station, and this could affect the localization. To avoid this situation, a mechanism is needed to detect this kind of failures. In order to detect link failures we use the concept of opportunistic routing. When a node transmits a packet, it starts a timer and listens to the channel. If it detects a packet sent by its interme- diate neighbor before the timer has expired, it considers that the transmission was successful.

A data packet has the following fields: source address, destination address, next hop, intermediate source address, hop count, uniqueID and RSS data, as shown in Fig. 2. The first is the address of the node which has measured the RSS. The desti- nation address is the address of the base station (typically zero). The next hop is the address of the intermediate neighbor stored by each node during the discovery phase. The intermediate node is the address of the node which sent the packet in the previous hop. The uniqueID field is used to avoid possible collisions. It con- tains a random value generated by the node that begins the communication toward the base and it does not change during the trip. Finally, the last field of the data packet contains the original RSS measurements.

```
 0                   1                   2                   3
 0 1 2 3 4 5 6 7 8 9 0 1 2 3 4 5 6 7 8 9 0 1 2 3 4 5 6 7 8 9 0 1
+-+-+-+-+-+-+-+-+-+-+-+-+-+-+-+-+-+-+-+-+-+-+-+-+-+-+-+-+-+-+-+-+
|          Source Address       |       Destination Address     |
+-+-+-+-+-+-+-+-+-+-+-+-+-+-+-+-+-+-+-+-+-+-+-+-+-+-+-+-+-+-+-+-+
|            Next Hop           |   Intermediate Source Address |
+-+-+-+-+-+-+-+-+-+-+-+-+-+-+-+-+-+-+-+-+-+-+-+-+-+-+-+-+-+-+-+-+
|           Hop Count           |           UniqueID            |
+-+-+-+-+-+-+-+-+-+-+-+-+-+-+-+-+-+-+-+-+-+-+-+-+-+-+-+-+-+-+-+-+
| RSS Data [0]  | RSS Data [1]  | RSS Data [2]  | RSS Data [3]  |
+-+-+-+-+-+-+-+-+-+-+-+-+-+-+-+-+-+-+-+-+-+-+-+-+-+-+-+-+-+-+-+-+
| RSS Data [4]  | RSS Data [5]  | RSS Data [6]  |RSS Data [....]||
+-+-+-+-+-+-+-+-+-+-+-+-+-+-+-+-+-+-+-+-+-+-+-+-+-+-+-+-+-+-+-+-+
```

Fig. 2 Structure of a *data packet*

The algorithm works in this way. Let A, B, C and D be four static nodes (see Fig. 3), with C being the base station. Node A measures the RSS of a packet sent by the mobile node, puts the data into a message, generates a random uniqueID, starts its timer and sends the packet to its intermediate neighbor, for example B. So the next hop address contained into the packet is B. This packet is received by both B and D. D checks the next hop field, and discards the packet, since it was di- rected to B. On the other hand, B checks the next hop field and accepts the mes- sage. Then it writes its address into the intermediate address field and the address of its next hop neighbor (C) into the next hop field, while the other fields are not changed. Then, it starts its timer and sends the packet to node C. C receives the packet, but due to the broadcast nature of the channel, A receives it too. If the timer of A has not yet expired, it checks if the intermediate address field is equal to its next hop neighbor (in this case B=B), in order to know whether the packet was generated by itself. Then it checks the uniqueID: if it is the one it generated before the transmission is considered successful. This uniqueID prevents possible collisions that could happen otherwise if two or more routes pass through the same node (imagine, for instance, a packet generated by a node E, following the route E-B-C, that is received by A).

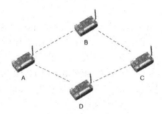

Fig. 3 Example of a small network

With this technique we consider a data packet as an acknowledgement. If a packet is not received before the timer expiration, we consider that the link has failed. Note that the base station has to send a packet too when it receives a message to avoid that the node sent the packet falling into the link failure case.

3.2.3 Link Failure

If a timer expires before the node receives the acknowledgement, we consider that the transmission has failed and that the link is unavailable. The node passes then into a *recovery mode*. In this case the node re-sends the packet, but introducing the broadcast address into the next-hop field. In this manner each node that receives the message will try to forward it to the base station using its own next-hop neighbor.

In this recovery mode the behavior can be configured to achieve a higher delivery rate. The node can restart its timer and repeat the retransmission several times. However one packet can be enough, especially if the network is dense and each node has several neighbors.

4 Experimental Evaluation

In this section we present the results of some experiments that were carried out to test the performance of the proposed algorithm in terms of energy consumption and communication delay. We implemented this algorithm in TinyOS 2.x for IRIS motes [15] and we deployed in our laboratory a small network, composed of ten static nodes, one mobile node and one base node connected to a computer.

The first experiment was aimed at evaluating the power consumption of the proposed algorithm and, in particular, at evaluating the energy reduction due to the use of the low-power MAC layer. To this end, we used the *BatteryC* module of TinyOS to measure the voltage level of the batteries of the nodes. We carried out one test running the proposed algorithm (with the LPL MAC layer) and then repeated it without the LPL MAC layer.

Fig. 4 shows the evolution of the voltage level (average voltage of the ten static nodes) in the two cases. The red line is referred to the case with LPL, while the blue one is referred to the test without the LPL MAC layer. From these data, and taking into account that the nodes stop working when the voltage goes below 2.2 V, we can estimate a battery life of 3 days without using the low power listening and 53 days with the low power listening MAC (with a new discovery phase each 10 seconds).

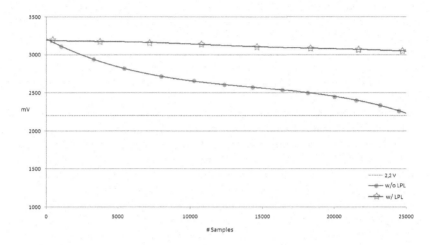

Fig. 4 Battery life w/ LPL and w/o LPL

The use of the LPL MAC layer reduces the energy consumption and increases the battery lifetime, but on the other hand, it introduces some delay in the multi-hop communications. According to the literature [100] the communication delay is equal to the product of the cycle duration and the length of the path. For example, as shown in Fig. 5, the delay to complete the path A-C-D should be twice a full cycle: At instant 1, A wants to communicate with C, so it begins sending the preamble, followed b the data. C wakes up at instant 3, and receives a small part of preamble (represented in black in the figure) and then the data (shown in green). At this point C wants to send this packet to D, so it begins sending the preamble at instant 4. D wakes up at 5, receives part of the preamble and then the packet. At instant 6 the communication has terminated. The delay is 2 full cycles (a full cycle is the time between the beginning of a white small square and the beginning of the next one).

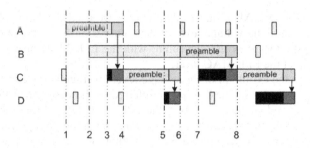

Fig. 5 Classic LPL behavior

In our implementation, this does not happen because the preambles in TinyOS are composed of many small messages. In this way, a mote can receive packets between a small message and the following. Since we use data packets in the preambles, the delivery is faster than in the classic case and delay is a random variable. In the worst case, the delay will be equal to the delay of the classic case, but in general, it will be lower and its value will depend on the load of the network.

Moreover, the TinyOS scheme enables mixed or interleaved communications, which are not possible with the classical scheme. For example, imagine that in Fig. 5, both A and B want to send a packet to D (through C): B wants to communicate with C at instant 2. Since the channel is busy it must wait until the channel is free, which happens only at instant 6. Then the communication B-C-D ends at the end of the green square of D. It does not matter which node (B or C) gets the channel at instant 4, the delay for the entire communication A-C-D and B-C-D is four full cycles. However, in our implementation the delays are much smaller (see Fig. 6), because at instant 2 (the instant is the same in Figure 5 and Figure 6), B can start sending data to C. At instant 4, C has received both packets from A and B, and at 5 both communications end. In this example, both transmissions A-C-D and B-C-D take only one full cycle to complete, but they could take more if there were others nodes or if the packets were bigger than 29 bytes (the maximum payload for a TinyOS packet).

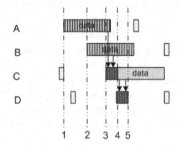

Fig. 6 TinyOS LPL behavior

Due to the behavior described above, the real delay in our implementation is smaller than in the classical case. We carried out a set of experiments to measure the delay in the discovery of a mobile node for different networks(2, 3 and 4-hop networks). A 2-hop network is composed by the base station, one static node and the mobile node; a 3-hop network has two static nodes and a 4-hop network has three static nodes. The delay was measured as the round trip time from the base station, to the mobile node and back to the base station. A total of 20 experiments were carried out for each type of network. Table 1 shows the average delay values obtained for the different types of networks.

Table 1 Average delay in milliseconds for 2, 3 and 4-hop networks.

2-hop	3-hop	4-hop
1065 ms	1797 ms	2563 ms

We can observe that the round trip delay increases 700-750 ms for each new hop. Note that in this test we set a full cycle time of 1000 ms, so in the classical LPL behavior the delays would be 4000 ms, 6000 ms and 8000 ms for 2, 3 and 4-hop network respectively.

5 Conclusions

In this paper we have presented a low power routing algorithm for localization-aware applications. The algorithm is based on the LPL MAC layer provided by TinyOS and the concept of opportunistic routing, and was conceived to provide a low-power mechanism to find the routes towards the base station for the RSS measurements collected by the nodes of the network.

We deployed a real network of ten nodes to test the performance of our algorithm. As shown in the experiments we obtained a battery life almost 20 times greater with the LPL MAC than without that MAC layer. Furthermore, the delay in the discovery of a node is lower than in classical duty cycle schemes due to the behavior of TinyOS LPL.

Further research should be focused on improving the routing performance, especially during the discovery phase avoiding the flooding. This could be done using the coordinates of the static nodes, since these values can be established during the network deployment.

Another aspect that can be further analyzed is the delay introduced by the duty cycling scheme. We are planning to evaluate the delay as a function of the path length, the hop count, the duty cycle and the network load in order to find the relations between these factors which could be useful when designing location-aware applications.

Acknowledgments. This work has been supported by the Government of Madrid under grant S2009/TIC-1485 (CONTEXTS) and by the Spanish Ministry of Science and Innovation under grant TIN2008-06742-C02-01.

References

1. Liu, H., Darabi, H., Banerjee, P., Liu, J.: Survey of wireless indoor positioningtechniques and systems. IEEE Transactions on Systems, Man and Cybernetics, Part C:Applications and Reviews 37(6), 1067–1080 (2007)
2. Gezici, S.: A survey on wireless position estimation. Wireless Personal Communications 44(3), 263–282 (2008)
3. IEEE 802.15.4 Standard, IEEE Standard for Information technology- Telecommunications and information exchange between systems - Local andmetropolitan area networks - Specific requirements. Part 15.4: Wireless Medium AccessControl (MAC) and Physical Layer (PHY) Specifications for Low-Rate WirelessPersonal Area Networks, WPANs (2006)
4. Lorincz, K., Welsh, M.: MoteTrack: a robust, decentralized approach to RF-basedlocation tracking. Personal and Ubiquitous Computing 11(6), 489–503 (2007)
5. Li, X.: RSS-based location estimation with unknown pathloss model. IEEE Transactions on Wireless Communications 5(12), 3626–3633 (2006)
6. Anastasi, G., Conti, M., Di Francesco, M., Passarella, A.: Energy conservationin wireless sensor networks: A survey. Ad Hoc Networks 7(3), 537–568 (2009)
7. Bachir, A., Dohler, M., Watteyne, T., Leung, K.: MAC essentials for wireless sensor networks. IEEE Communications Surveys & Tutorials (2010)
8. TinyOS Community Forum, http://www.tinyos.net
9. Polastre, J., Hill, J., Culler, D.: Versatile low power media access for wireless sensor networks. In: SenSys 2004: Proceedings of the 2nd International Conference on Embedded Networked Sensor System, New York, pp. 95–107 (2004)
10. Jurdak, R., Baldi, P., Videira Lopes, C.: Energy-aware adaptive low power listening for sensor networks. In: Proc. 2nd International Workshop for Networked Sensing Systems, San Diego (2005)
11. Buettner, M., Yee, G., Anderson, E., Han, R.: X-MAC: A Short Preamble MAC Protocol For Duty-Cycled Wireless Networks. Technical Report CU-CS-1008-06, University of Colorado at Boulder (2006)
12. TinyOS 2.x Contributing Code,
    http://tinyos.cvs.sourceforge.net/viewvc/tinyos/
    tinyos-2.x-contrib/
13. AODV Routing RFC, http://www.ietf.org/rfc/rfc3561.txt
14. Liu, H., Zhang, B., Mouftah, H., Shen, X., Ma, J.: Opportunistic routing for wireless ad hoc and sensor networks: Present and future directions. IEEE Communications Magazine 47(12), 103–109 (2009)
15. IRIS Datasheet, http://www.memsic.com/products/
    wireless-sensornetworks/wireless-modules.html

Ambient Intelligence Acceptable by the Elderly

Rational Choice Theory Model of Technology Evaluation for Deep Design

David Zejda

Abstract. Ambient systems may support the elderly in many aspects of their lives, bringing new level of comfort, higher safety, and better health. But, as we revealed in parallel research on deep design, there are dissonances in what do the elderly wish and what the intelligent technologies indeed offer. The dissonances may lead to reluctant acceptance or even to rejection of possibly beneficial products. In this paper we borrowed concepts from economics and transformed them into generic model which captures the mental process of evaluation of new aspects of life. Proposed model is specifically aimed on ambient intelligence products, viewed from the eyes of possible elderly users, though not necessarily limited to this particular focus. The process of evaluation, leading either to acceptance or to rejection, may be described as a sequence of rational selections from available options, based on perceived benefit (utility) and cost (in terms of time, effort, support from others). Simplifying preconditions introduced in the paper reduce the model to optimization problem of linear programming. In conclusions we discuss limitations of the model and suggest further possible refinements and evaluation.

Keywords: Ambient intelligence, acceptability, evaluation, rational choice, the elderly, linear programming.

1 Introduction

Idea of human-centric design is not entirely new. E.g. Norman [1] focused on the role of emotions in our cognition, Callejas et al. [2] suggested to include users in early phases of design, Vogiazou et al. [3] introduced a 'design for emergence', where users are being observed in their daily activities, technology probes promoted e.g. by Hutchinson et al. [4] may get close to users and Jordan [5] refused common task-centric design as dehumanizing. Actually, we hope, dream, sorrow,

---

David Zejda
University of Hradec Králové, Faculty of Informatics and Management, Czech Republic
e-mail: david.zejda@uhk.cz

J.M. Molina et al. (Eds.): User-Centric Technologies and Applications, AISC 94, pp. 103–110.
springerlink.com   © Springer-Verlag Berlin Heidelberg 2011

fear, desire and aspire, and that make us human. With deep design, the innermost sources of affinity which people feel to certain object or service are being targeted. Successful products have to pass all phases of evaluation and appropriation – attract attention first, show rewards soon, be appropriated into habitual usage finally. Failure in any of the steps effectively means refusal of the product. [6] High-order reinforcers stemming from the deep needs constitute the power which causes, whether certain product will succeed use in the long-term. [7]

If we wish to overcome scepticism, qualms, and anxiety which many older people feel towards technology and motivate them to learn new things, it is even more important to target their needs as precisely as possible. According to [8] the elderly in general do not wish to be monitored, but they wish to keep their independence, they wish to live in their home, to stay in touch with their close, to feel competent and helpful and dignified. Current sensoric monitoring systems (fall or other crisis detection) or cognitive support systems (e.g. reminders pushing users to take medication timely) actually focus on care-givers more than on care-receivers themselves. [9] To deliver technologies acceptable by the elderly and perceived by them as improvement to their lives, we should hold back of technology and put ourselves in their pattern of thinking [6]. We already followed the path of research and concluded in formulation of deep design approach [8], which emphasizes revealed deep needs of users as the most important foundation for subsequent design.

Once we revealed the deep needs, we may examine the problem of acceptability and mental processes behind from various perspectives, stressing psychological, sociological, technical or health aspects. Each of the approaches may be beneficial, but if we wish to achieve a model formal enough to allow calculations, good way is to go out of economy. Gary Becker proved that the approach may be fruitful. He captured various aspects of human life and behaviour [10] including discrimination [11], crime and punishment [12], addiction [13], beggary and compassion [14], human capital [15], love, marriage and family [16] into rational choice theory, which transcends narrow borders of pure economy. According to the theory, nearly all human behaviour may be explained as rational reasoning. Rational person decides as if balancing costs against benefits. [17] Person's aim is to maximize his advantage.

The general model of evaluation introduced in the paper describes early phases of evaluation, when somebody is thinking whether to try certain product or not, with particular focus on ambient and similar technologies evaluated by elderly users. The model should help to answer questions, such as: How elderly users perceive ambient technologies aimed to improve their lives? Which steps may be identified in the process of the evaluation? How do they compare alternatives if there are any? What causes that the technologies are being refused? While in economy advantage is being measured with money, rational choice theory emphasizes more subtle determinants of comfort, such as self-worthiness and social relations (Bentham [18] or Marshall [19]), which closely resonates with our deep design approach, focused on inner feelings. Similarly, cost in our model instead of its monetary representation falls into range of softer expenditures – effort, time, external support. Model is defined generally, as a foundation for further formalization and further concretization in terms of particular shapes of introduced functions and their coefficients. The concretization should be based on statistical research.

2 The Utility

Both in economy and rational choice theory, utility acts as a measure of perceived benefit. Personally unique utility (payoff) function reflects personal preferences. Actions driven by the utility function are constrained by budget, abilities, time available. The same basic idea is behind the proposed model of evaluation. Our research on attitude of the elderly towards intelligent technologies [8] resulted in four main clusters of deep needs of the elderly. Let's call them *sources of comfort* A .. D (or comfort sources) in our model. Each source of comfort has finite real number assigned, reflecting perceived level of fulfilment of the source by certain person's life situation:

A .. social touch
B .. autonomy with anticipated support
C .. feeling of being competent
D .. feeling of helpfulness and self-worth

Let's have a *utility function* U, which reflects personal preferences on importance of the sources of comfort. The function assigns total utility value to each combination of comfort sources:

$$U = U(A, B, C, D) \tag{1}$$

Simple form of the utility function assigns weight to each of the comfort sources. Let's have *aspects of life* i_1 .. i_n (or life aspects) as a complete set of determinants of comfort sources of size n. Aspects of life may be e.g. "living with the family", "walking daily", "having a telephone" or "having a telephone call every day". Each aspect of person's life may influence each of the sources of comfort. The influence of aspects of life on comfort sources may be captured as a set of *influence vectors*:

$$
\begin{aligned}
i_1 &= [a_1, b_1, c_1, d_1] \\
i_2 &= [a_2, b_2, c_2, d_2] \\
&\quad\cdots \\
i_n &= [a_n, b_n, c_n, d_n]
\end{aligned}
\tag{2}
$$

Influence of certain aspect of life on certain source of comfort (e.g. b_2, the influence of aspect i_2 on comfort source B) may by either positive (raising the comfort source) or negative (lowering the comfort source). Total level of each comfort source is fully determined by influence of all aspects of person's life on the comfort source:

$$
\begin{aligned}
A &= \text{sum}(a_1 .. a_n) \\
&\quad\cdots \\
D &= \text{sum}(d_1 .. d_n)
\end{aligned}
\tag{3}
$$

According to the utility function, each aspect influences overall utility. As well as utility function, effects of life aspects are subjective. E.g. though many people perceive living with their family as highly beneficial because it strongly increases level of social touch, some may prefer living alone, because they feel more competent this way and the positive effect of higher competency outweighs lowered social touch in their case. Both assigned weights and utility functions play roles in the evaluation. Changes in the set of aspects of life may capture dynamics of changes in person's life. Introducing a new aspect of life (e.g. new cell phone) adds relevant element (line) to the set of aspects in the model. Similarly, replacing one aspect for another (weekly visit of children changed for daily phone calls) removes one aspect from the model and adds another.

3 First Phase of Technology Evaluation: "Is It Beneficial?"

The first step in the process is evaluation is reasoning on anticipated benefits, constraints (cost, time, etc.) do not matter. Adoption of certain life aspect influences total utility, dU_n is the *utility difference on adopting aspect* i_n:

$$dU_n = U(A+a_n, B+b_n, C+c_n, D+d_n) - U(A, B, C, D) \qquad (4)$$

In the equation A .. D are levels of comfort sources before introducing the new technology and a_n .. d_n are influences of the new aspect. To pass the first stage of evaluation, evaluated aspect (e.g. new product introduced into life) has to influence the total utility positively. Anticipated dU_n must be higher than zero, otherwise it will be rational to refuse the aspect i_n. Seemingly paradoxical situation, when certain technology is being refused by the elderly even though they do not need to pay single penny, exhibit significant effort or sacrifice time, may be explained as a failure in the first evaluation stage, which we call *refusal of the first kind*.

4 Second Phase of Technology Evaluation: "Is It Reachable?"

Multitude of life aspects with positive influence on total utility is available, but not all of them are reachable. Reachability of life aspects is limited by certain constraints. Economy emphasizes monetary dimension of the constraints, calling them "budget". We in the contrary assume, that cost does not play role at all in the way how the elderly evaluate aspects of life. The assumption reflects situations where those who evaluate a product (the elderly) are not those who bear relevant financial expenses, someone else, either children or an institution, is the one who has to pay. Also, the model reflects situations where cost has been paid already (e.g. somebody already bought a gift, presentee is about to evaluate). As more relevant to our scenario we take another constraints called *resources* into account, though could be easily put back, if necessary:

T .. time (time necessary to manage/appropriate/perform the aspect of life)
E .. effort (effort and abilities necessary)
S .. support by others (help from family, care givers and others)

In our model we presume, that all the resources are limited. Each person has only certain time available for all his activities (24 hours daily, or less if we deduct the time for necessities), certain amount of power, determination, intellectual and mental skills etc. (involved in compound "effort" resource), and may ask certain level of support from others (determined by their willingness in the case of family members, and e.g. by received pension in the case of care givers – introducing cost indirectly). Let's have *exert function* f_n, which describes which amounts of resources (t_n, e_n, s_n) are necessary to manage, appropriate and/or perform aspect of life i_n:

$$f_n = f_n(t_n, e_n, s_n) \tag{5}$$

Aspect reachability is binary in our model. With given combination of resources, each aspect is either reachable (and possibly accepted) as a whole, or not reachable; aspect can be accepted neither only partially, nor more than fully. So, f_n may return value 1 or 0, where $f_n = 1$ means that aspect i_n is reachable exerting the given combination and $f_n = 0$ means aspect i_n is not reachable. If certain aspect is not reachable (requires more time, effort or support than available), it can't be accepted though perceived as beneficial, which leads to *refusal of the second kind*.

5 Third Phase of Technology Evaluation: "Is It the Best Choice?"

Even if beneficial and reachable still we do not know enough to say whether an aspect will be accepted. If resources are limited, rationally deciding human will compare all available choices to select the most beneficial combination available. The third phase of evaluation takes into account anticipated benefits, resources available, and resource intensity of life aspects. In reality, resources may be substituted, e.g. insufficient abilities may be compensated with increased help from others or with more time spent. But to keep the model simple, for further reasoning we presume incommutable resources in exert function. It means that there are certain externally determined optimal input proportions given for each person's aspect of life. And because exert function is binary, it would be useless to add more resources above the optimal (minimal sufficient) levels.

So, let's have three available aspects of life i_1, i_2, i_3 and we wish to compare them in the context of available resources to choose the optimal combination. The aspects of life have potential to influence total utility by dU_1, dU_2, dU_3 respectively according to subjective utility function. If $[t_1, e_1, s_1]$, $[t_2, e_2, s_2]$, $[t_3, e_3, s_3]$ are vectors of optimal resource levels necessary to adopt the aspects, corresponding exert functions are defined as:

$$f_1(t >= t_1 \wedge e >= e_1 \wedge s >= s_1) = 1; f_1(t <t_1 \vee e < e_1 \vee s < s_1) = 0$$
$$f_2(t >= t_2 \wedge e >= e_2 \wedge s >= s_2) = 1; f_2(t <t_2 \vee e < e_2 \vee s < s_2) = 0 \qquad (6)$$
$$f_3(t >= t_3 \wedge e >= e_3 \wedge s >= s_3) = 1; f_3(t <t_3 \vee e < e_3 \vee s < s_3) = 0$$

The goal of rationally reasoning human is to maximize total U within borders of reachable opportunities (defined with levels of available resources T, E, S). He asks which aspects of life (out of available i_1, i_2, i_3) should be adopted. Adoption of aspect of life i_n is expressed as variable x_n, which may take value 1 or 0 (the aspect is being adopted and aspect is not being adopted respectively). The model may be expressed as follows:

$$dU = x_1 dU_1 + x_2 dU_2 + x_3 dU_3$$
$$\max(dU) \text{ on condition}$$
$$T >= t_1 x_1 + t_2 x_2 + t_3 x_3 \qquad (7)$$
$$E >= e_1 x_1 + e_2 x_2 + e_3 x_3$$
$$S >= s_1 x_1 + s_2 x_2 + s_3 x_3$$

The model may be solved as optimization problem of linear programming. Aspects of life which do not pass the third stage of evaluation suffer from the *refusal of the third kind*.

6 Conclusions

In the paper we applied rational choice theory methods on the problem of acceptability of intelligent ambient technologies by the elderly with aim to formalize possible causes of refusal. According to the findings, product has to pass three levels of rational evaluation to be accepted and in contrary three kinds of refusal may occur, refusal of the first kind if a product is not beneficial, refusal of the second kind if a product is not reachable and refusal of the third kind if a product does not belong to the most beneficial reachable set of aspects. General model of acceptance introduced in the paper reflects the evaluation process. Intentionally we designed the model as concise, simple and easily understandable rather than highly formal. Arbitrary simplifying presumptions including binary exert function (aspect of life is either reachable or not), and limited and incommutable resources (time, effort and help can't be substituted) allowed to express the model in the form of optimization problem of linear programming. To reflect reality better, the model could be refined, e.g. the presumption of incommutable resources could be released. On the other hand, for practical application further simplifications of the model would be necessary. From the virtually infinite set of aspects of life only certain life aspects would have to be selected, either the most influential or the most relevant to the area of interest. Statistical research will be necessary to find appropriate shapes of utility and exert functions including function coefficients. Statistical evaluation may help also to examine how closely the model matches the real process of evaluation. The model may be proclaimed as inappropriate or useless based on statistical data if we find out, that it is not possible to concretize e.g. influence of aspects of life on sources of comfort.

Despite the inaccuracy caused by simplifications, socio-economic and psycho-economic models may help to grasp processes in our minds and may help e.g. to reveal weaknesses of technologies and products. Parallels of our model with economics suggest interesting analogies, such as income and substitution effects as an explanation how changes in requirements on resources influence acceptance of *other* aspects of life. With the help of the models we may not only bring more acceptable products, but also deepen our understanding of each other, and even understanding of ourselves. Regardless to any possible refinements, it is necessary to interpret the models with high caution. Human mind is highly complex system and any model brings only very rough insight.

Acknowledgments. The research was supported by grant UHK FIM specific research 2110/2010 – "Intelligent Social Technologies for Quality Life of Elderlies."

References

[1] Norman, D.A.: Emotional Design: Why We Love (or Hate) Everyday Things, 1st edn. Basic Books, New York (2003)
[2] Callejas, Z., López-Cózar, R.: Designing smart home interfaces for the elderly. SIGACCESS Access. Comput. 95, 10–16 (2009)
[3] Vogiazou, Y., Reid, J., Raijmakers, B., Eisenstadt, M.: A research process for designing ubiquitous social experiences. In: Proceedings of the 4th Nordic Conference on Human-Computer Interaction: Changing Roles, pp. 86–95 (2006)
[4] Hutchinson, H., et al.: Technology probes: inspiring design for and with families. In: Proceedings of the SIGCHI Conference on Human Factors in Computing Systems, pp. 17–24 (2003)
[5] Jordan, P.W.: Designing Pleasurable Products. CRC Press, Boca Raton (2002)
[6] Veldhoven, E.R., Vastenburg, M.H., Keyson, D.V.: Designing an Interactive Messaging and Reminder Display for Elderly. In: Aarts, E., Crowley, J.L., de Ruyter, B., Gerhäuser, H., Pflaum, A., Schmidt, J., Wichert, R. (eds.) AmI 2008. LNCS, vol. 5355, pp. 126–140. Springer, Heidelberg (2008)
[7] Carroll, J., Howard, S., Vetere, F., Peck, J., Murphy, J.: Just What Do the Youth of Today Want? Technology Appropriation by Young People. In: Proceedings of the 35th Annual Hawaii International Conference on System Sciences (HICSS 2002), vol. 5, p. 131.2 (2002)
[8] Zejda, D.: Deep Design for Ambient Intelligence: Toward Acceptable Appliances for Higher Quality of Life of the Elderly. In: 2010 Sixth International Conference on Intelligent Environments, pp. 277–282 (2010)
[9] Miyajima, A., Itoh, Y., Itoh, M., Watanabe, T.: "Tsunagari-kan" Communication: Design of a New Telecommunication Environment and a Field Test with Family Members Living Apart. International Journal of Human-Computer Interaction 19(2), 253 (2005)
[10] Becker, G.S.: The economic approach to human behavior. University of Chicago Press, Chicago (1976)
[11] Becker, G.S.: The economics of discrimination. University of Chicago Press, Chicago (1971)

[12] Becker, G.S.: Crime and punishment: An economic approach. Journal of Political Economy 76(2) (1968)
[13] Becker, G.S., Murphy, K.M.: A theory of rational addiction. The Journal of Political Economy 96(4) (1988)
[14] Becker, G.S.: Accounting for tastes. Harvard Univ. Pr., Cambridge (1998)
[15] Becker, G.S., et al.: Human capital. National Bureau of economic research, New York (1975)
[16] Becker, G.S.: A Treatise on the Family. Harvard Univ. Pr., Cambridge (1991)
[17] Friedman, M.: Essays in positive economics. University of Chicago Press, Chicago (1953)
[18] Bentham, J.: Theory of legislation. Adamant Media Corporation (1876)
[19] Marshall, A.: Principles of economics: an introductory volume (1920)

Radio Emitters Location through the Use of Matrix-Pencil Super-Resolution Algorithm

Raúl González-Pacheco* and Felipe Cátedra**

Abstract. This article focuses on the location of radiating sources using calculations encompassing wide ranges of an electromagnetic signal. The purpose of using the Matrix Pencil algorithm is to consider the distance at which the transmitting antennas are, implementing a measurement procedure throughout simulations. In order to do these simulations we use a tool whose algorithm is based on the UTD (Uniform Theory of Diffraction) and/or ray tracing and that determines the value of the real surroundings electromagnetic field under study and that is validated to provide trustworthy results in a reduced runtime. Thus the signal spatial signature is observed. A new and simple approach for estimating distances through Matrix Pencil is presented. The technique discussed in this article proceeds from using the Matrix Pencil super-resolution algorithm and vary only, but importantly, in how this sum of complex exponentials are use in achieving the parameter estimation.

1 Introduction

In recent times the methodology of approximating a function by a sum of complex exponentials has found applications in other areas of electromagnetic, i.e. in antenna-pattern synthesis; extraction of the s-parameters of microwave-integrated circuits; in the analysis of signal propagation over perforated ground planes; in the computation of input impedance of electrically wide slot antennas; in the analysis of complex modes in lossless closed conducting structures; in multiple transient signal processing; in inverse synthetic-aperture radar; in high-resolution moving targets imaging and in radio-location finding.

Talking about radar, sonar, radio astronomy and seismology, one of the important problems to be solved is the spatial source location through passive sensors.

Raúl González-Pacheco · Felipe Cátedra
Dpto. Ciencias de la Computación, Universidad de Alcalá
28806 Alcalá de Henares. Madrid. Spain
* Fax: +34 91 7461607
e-mail: `rgonzalez@iberia.es`
** Fax: +34 91 8856646
e-mail: `felipe.catedra@uah.es`

J.M. Molina et al. (Eds.): User-Centric Technologies and Applications, AISC 94, pp. 111–120.
springerlink.com   © Springer-Verlag Berlin Heidelberg 2011

In essence, a radio location system can operate measuring and processing physical amounts related to the radio signals travelling between a mobile terminal (MT) and a set of base stations (BSs), like the time of arrival (ToA), the angle of arrival (AoA) or the signal strength [2].

The use of the signal strength magnitude is based on the fact that the average power of a radio signal decays over distance, following a well-known law.

Many authors discussed the generalized eigenvalues problems in matrix theory, both from the algebraic point of view (see [3] and [4] and the references therein) and from the numerical (computational) point of view (see [5] and [6], and the references therein).

The classic method for searching of spatial signatures is based on the direct application of the Fourier transform. Its limitation is that it is not possible to distinguish between sufficiently close signals. This gave way to the great resolution methods, whose philosophy is based on the underlying signal model, that is, they assume that the data are adjusted to a model which parameters contain the information to be considered. The models based on analysis methods increase the spectrum resolution.

The purpose of this paper is to find a new and simple approach for calculating the location of radio emitters through generalized eigenvalues and its corresponding eigenvectors of a regular Matrix Pencil. The approach is given by transforming the generalized eigenvalues problem of a regular Matrix Pencil into a usual eigenvalues problem.

An important research effort in the development of a solution that uses only one base station has been undertaken.

In the scenario presented in this paper, the energy contained in the field of waves, created by the emitter, is gathered by a sensor. Later a receiving system measures the response of this sensor. A flat faceted model is used to represent the propagation scenario.

The paper is organized as follows: point 2 describes the signal model, leaving for section 3 the comparison between polynomial and Matrix-Pencil methods. Point 4 presents the testing scenario. The results of the simulations are inserted at point 5 and, finally, point 6 details the conclusions.

2 Signal Model

The signal model of the observed late time of electromagnetic energy scattered response from an object can be formulated as:

$$y(t) = x(t) + n(t) \approx \sum_{i=1}^{M} R_i \exp(s_i t) + n(t)$$

$$0 \leq t \leq T$$

where:

  $y(t)$= observed time response
  $n(t)$= noise in the system

$x(t)$= signal
R_i= residues or complex amplitudes
s_i= $-\alpha_i + w_i$
α_i= damping factors
w_i= angular frequencies ($w_i = 2\pi f_i$)

Beginning with the coverage analysis, in order to lead the study to the vectorial impulse / spatial signature answer, a coverage calculation program is used. The selected tool for outdoors and indoors is NEWFASANT [7]. The code is based on the uniform version of the Geometric Theory of Diffraction, and takes into account the effects from direct, reflected, diffracted in the edges, double reflected, reflect-refracted, diffracted-reflected fields as well as the influence of the ground.

The Matrix Pencil algorithm will determine the argument of the exponential, i.e. each of the M different arguments that appear in the equation.

The signal synthesized as a sum of complex exponentials (to be determined), whose argument varies linearly with the number of samples (i in the formula above) is collected by a sensor, taking as input variables the frequency range and the number of samples. The output gives the vector components for each ray. We work with a vector magnitude, however, Matrix Pencil applies to scalar magnitudes; therefore we work with only one of the field components, any of them.

It has to be checked that the absolute value of the component x, y or z is approximately the same for all frequencies. If not, the increase in frequency or the number of them should be reduced.

It dispenses with noise, because Matrix Pencil does not require it.

The key issue of Matrix Pencil is the estimation of the number of arriving waves. From the study of the eigenvalues magnitude of the problem of eigenvalues which returns the Matrix Pencil method, it turns out that the number of significant eigenvalues is an estimation of the number of arriving waves. Nontrivial eigenvectors of distinct eigenvalues are the most useful ones because they possess the desired properties for channel estimation. When an eigenvalue is said to be distinct, it does not have any multiplicity and has only one corresponding eigenvector.

If the amplitude of the waves is similar, the eigenvalues clear easily from the noise level, but if this is not the case, it might confuse us and cannot correctly estimate the number of arriving waves. To have a good estimation of this amount is capital, since in the Matrix Pencil algorithm matrices are truncated and manipulated taking this value into account. If we wrong the number of waves and we force to detect a wrong number of them, the resulting frequencies would be an average of which actually exist.

Automation of the eigenvalues search demanded a major research effort.
The response of the Matrix Pencil algorithm is the z values,

$$z_n = e^{-j\frac{2\pi}{c}\cdot\Delta f \cdot D_n}$$

where:

$z_n=$ MP´s solution for the n wave
D_n = distance travelled by the n ray

but do not appear in order, this means that we do not know what ray is each.

3 Relationship between the Polynomial and Matrix Pencil Methods

The term "Pencil" is originated with Gantmacher [8] Even though the two methods were originated from different approaches; there is a link between the two, as is shown below.

It can be shown that the roots of the polynomial $\sum_{k=0}^{L} a_0 z^{-k}$ (with $a_0=1$), are the eigenvalues of the matrix:

$$[C_1] = \begin{bmatrix} 0 & 0 & \dots & 0 & -a_L \\ 1 & 0 & \dots & 0 & -a_{L-1} \\ 0 & 1 & \dots & 0 & -a_{L-2} \\ . & . & . & . & . \\ . & . & . & . & . \\ . & . & . & . & . \\ 0 & 0 & \dots & 1 & -a_1 \end{bmatrix}_{LxL} = [U_2, U_3, \dots, U_L, Y_1^+ y]$$

where U_i is the (Lx1) vector with the ith element equal to 1 and all other elements zero.

$[Y_1]^+[Y_2]$ can be written as:

$$[C_2] = [Y_1]^+[Y_2] = [Y_1^+ y_{-1}, Y_1^+ y_{-2}, \dots, Y_1^+ y_{-L}]$$

with

$$y_{-k} = [y_k, \dots, y_{k+N-L-1}]$$

where K=1,…,L

As we can see, the ith column of $[C_2]$ is a solution of the following equation:

$$Y_1[b] = [y_{-i}] \qquad (1)$$

But in $[C_1]$, only the last column vector is the minimum-norm solution with $i=L$, while in $[C_2]$, all column vectors are minimum-norm solutions of equation (1) $[C_1]$ and $[C_2]$ are identical if $L=1$. Note that $[C_1]$ has M signal eigenvalues at z_i, and

L-M extraneous eigenvalues that are non-zero and located inside the unit circle, while [C_2] has M signal eigenvalues at z_i, and L-M extraneous eigenvalues that are zero. Therefore, we see that the polynomial method and the Matrix Pencil are different if $L>M$, and identical if $L=M$. For the over determined case, $L>M$, the results are different due to the use of different numerical recipes, even though the input data are identical.

The results of the two methods are different under noise. It can be shown that under noise, the statistical variance of the poles z_i for the Matrix Pencil method is always less than that of the polynomial method.

With equality when $L=M=1$. From $M \geq 2$, the two methods yield different variances. The Matrix Pencil method does deteriorate when the signal-to-noise ratio (SNR) decreases below about 25dB, unless one utilizes the band pass version of the Matrix Pencil method (BPMP).

4 Testing Scenario

The formula below refers to the sample i^{th} and assumes the no existence of noise. It seeks to characterize the argument of the complex exponentials (M in this case)

$$x_i = \sum_{l=1}^{M} \mid h_l \mid e^{\left((\alpha_l + j\omega_l)(i-1) + \phi_l \right)} \qquad i=1 \cdots N$$

To estimate the M frequencies $\omega_i \quad i = 1, \cdots, M$ (assuming that α_i is equal to zero, i.e. complex exponentials without real part) that define the signal vector $\mathbf{x}$, we must solve the following problem of generalized eigenvalues / eigenvectors:

$$\left(G_2 - \xi G_1 \right) \boldsymbol{p} = \boldsymbol{0}$$

Then G_2 and G_1 arrays are generated.

Next step is to find the eigenvectors ξ_i (one for each detected signal) and then the relationship between eigenvectors and the arguments of the exponentials are focused)

What you get with Matrix Pencil is $e^{j\omega_1}, e^{j\omega_2}, \cdots e^{j\omega_N}$, since $\xi_i = e^{j\omega_i}$, being $\xi_1 = e^{j\omega_1}$.

Frecuenciaw1=-(imag(log(ξ_1)));

whereω_l is the exponential argument except for the term (i-1).

It is hoped that the first eigenvalue is the match for the direct beam (which leads to a greater eigenvalue) The less distance shall correspond to the direct beam. The use of the pseudo inverse determines the amplitude, associating the greater term to the direct beam.

The description that follows, assumes that two reflected beams reach the observation point. To include more effects is immediate.

The distance that separates the source to the point of observation is D1; the traveling wave suffers reflection by the surface A, running a distance D11. In the model to consider a second reflection in area B is taken into account as well, which gives rise to the D21 distance) The figure to be determined is D1.

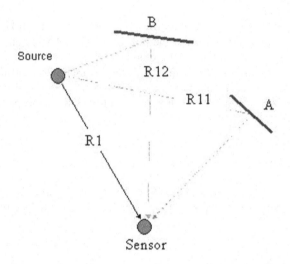

Fig. 1 Schematic source – sensor position

To do this, multiple FASPRO simulations taking 32 frequencies, 1 MHz equal-ly moved away, are performed. As shown, these 32 frequencies will play the role of 32 sensors.

The aim is to transform the MP estimation into distances. We shall see, MP gives the values of D1, besides D11 and D12.

The frequency of the ith wave is:

$$f_i = f_0 + \Delta f \cdot (i-1)$$

Then the signal gathered for this frequency (which is what you get from FASPRO) will follow this model:

$$E_i = E_i e^{\frac{-j2\pi}{c} \cdot f_0} \left(e^{-j\frac{2\pi}{c} \cdot \Delta f \cdot D_1 \cdot (i-1)} + R_A e^{-j\frac{2\pi}{c} \cdot \Delta f \cdot D_{11} \cdot (i-1)} + R_B e^{-j\frac{2\pi}{c} \cdot \Delta f \cdot D_{12} \cdot (i-1)} \right)$$

being E_i the incident field and R_A and R_B the coefficient of reflection in areas A and B.

Matrix Pencil assumes that the amplitude of arrival waves is the same for all values of i, so it checks that, for these frequency hopping, the amplitude varies least.

5 Results

In order to get results, different environments in which produce multipath with different lengths and angles of arrival were simulated.

All along the simulations it was verified that the excessive proximity of two complex exponentials to be detected, creates difficulties to Matrix Pencil.

Simulations were made on the basis of sweep frequencies from an initial frequency to another final. In this set of frequencies the amplitude which gives the GTD program assured the signal model was the right one, that is, that in the received field only changes were due to the phase term.

Fig. 2 illustrates the effectiveness of the Matrix Pencil algorithm. It is able to generate an accurate estimate when the waves are not very close together.

On the other hand, Matrix Pencil is severely influenced by the proximity of the complex exponentials to be detected. When they are much approximated, the response has a significant diversion.

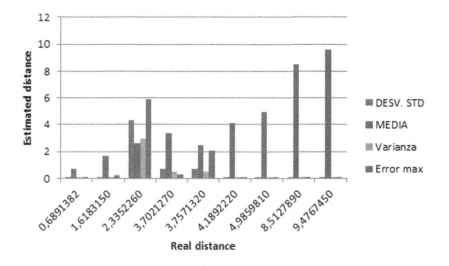

Fig. 2 Effectiveness of the Matrix Pencil algorithm

Fig. 3 shows the changes in the standard deviation based on the number of frequencies, the sample frequency interval and the Matrix Pencil parameter. In general standard deviation decreases with the Matrix Pencil parameter and increases with the interval between frequencies.

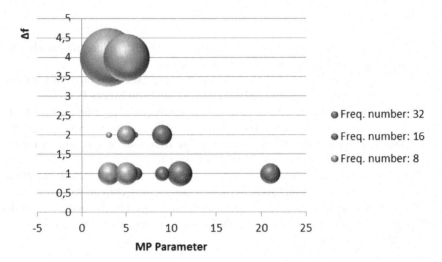

Fig. 3 Standard deviation changes

Fig. 4 shows the behavior of the variance. This strongly depends on the number of frequencies of the sample as well as the interval between these frequencies. Decreases with the number of frequencies and when the interval between frequencies gets smaller.

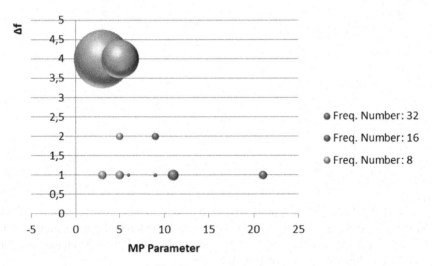

Fig. 4 Variance behavior

Fig 5 represents the maximum errors obtained all along the simulation process. It is clearly exposed the problem Matrix Pencil has when it works with very close exponential complex. However, under normal conditions, the error was below 1.9%.

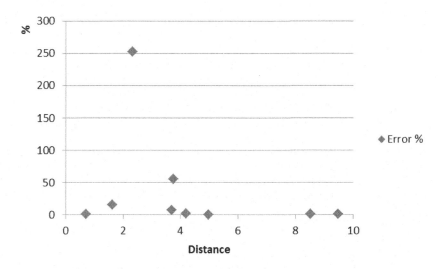

Fig. 5 Maximum errors

6 Conclusions

A novel emitter location Matrix Pencil algorithm is investigated in this paper. The power of this estimator lies on the fact that Matrix Pencil algorithm is very efficient and works well even with a single snapshot. It becomes an ideal choice for real time applications over other algorithms as they usually trade speed for accuracy or vice versa.

From previous research conclusions [8], Matrix Pencil was shown to be a more effective technique than MUSIC. It is better than what MUSIC algorithm can achieve. Only one snapshot is required for Matrix Pencil to accurately identify the length of the multipath components. The ability to estimate the length of all multipath components may allow in the development of new applications such as the RAKE receiver.

Matrix Pencil method provides smaller variance of the parameters than other methods like polynomial and it behaves very well in presence of noise as well.

Simulation is virtually done in real time, since we are dealing with a very fast algorithm. That means this version of Matrix Pencil can be implemented in hardware utilizing DSP chips operating in real time.

Acknowledgments. This work has been supported in part by the Comunidad de Madrid Project S-2009/TIC1485 and by the Castilla-La Mancha Project PPII10-0192-008, Technology Projects TEC 2007-66164 and CONSOLIDER-INGENIO Nº CSD-2008-0068.

References

[1] Sarkar, T., Pereira, O.: Using the Matrix-Pencil Method to Estimate the Parameters of a Sum of Complex Exponentials. Syracuse University, New York (1995)

[2] Messier, G., Fattouche, M., Petersen, B.: Locating an IS-95 Mobile Using Its Signal. In: Conf. Rec. The Tenth International Conference on Wireless Communications (Wireless 1998), Calgary, AB, Canada, vol. II, pp. 562–574 (1998)

[3] Gantmacher, F.R.: The Theory of Matrices. Chelsea Publishing Company, New York (1959)

[4] Benner, P.: An Arithmetic for Matrix Pencils: Theory and New Algorithms, Fakultät für Mathematik, Berlin (April 2004)

[5] Golub, G.H., Ye, Q.: An inverse free preconditioned Krylov subspace method for symmetric generalized eigenvalue problems. SIAM J. Sci. Comput. 24(1), 312–334 (2002)

[6] Wilkinson, J.H.: The Algebra Eigenvalue problems. Clarendon, Oxford (1965)

[7] http://www.fasant.com

[8] González-Pacheco, R., Cátedra, F.: Location Based Services: a new era for the use of super-resolution algorithms. In: Ubiquitous Computing & Ambient Intelligence, Salamanca (2008)

Author Index